Our Inevitable Third Spinoza Controversy

Katherine C. Snow

Our Inevitable Third Spinoza Controversy

Ontologies of Necessity and Environmental Nihilism

palgrave
macmillan

Katherine C. Snow
Princeton University
Princeton, NJ, USA

ISBN 978-3-032-03044-3 ISBN 978-3-032-03045-0 (eBook)
https://doi.org/10.1007/978-3-032-03045-0

This Palgrave Macmillan imprint is published by the registered company Springer Nature Switzerland AG.
The registered company address is: Gewerbestrasse 11, 6330 Cham, Switzerland

To a future chimera society
sic semper anthropos

Competing Interests The author has no competing interests to declare that are relevant to the content of this manuscript.

Contents

CHAPTER 1

The First Spinoza Controversy: *Nihil* or Chimera

The end of the eighteenth century in what would later become Germany saw divisions among Pietists, Catholics, Lutherans, Moravians, and a host of other sects; the French Revolution – which in Safranski's view showed that "thinking and writing not only interpreted but also changed the world … that it was only a matter of finding the right thoughts[1]" – shook the landscape as well, with both the promise and the threat of the arrival of a new age of European thought with potentially stark social consequences.

The turns eventually taken in Western thought during this period, including the enthusiastic exploration by the early German romantics of a new view of nature as a whole, would in fact have long-running ramifications which went far beyond contributing to dramatic political and social change. These deeper ramifications are rarely examined by philosophers or historians today, most of whom at least tacitly accept and utilize a German romantic-influenced view of nature (a version of which was also later taken up by modern science). In order to fully grasp any problems inherent in the West's romantic turn, including the perhaps-counterintuitive reasons this turn has inevitably driven the West's dismantling of the natural world (and its replacement with the domesticated, the synthetic, and the informatic), we must begin by returning to the period when this romantic turn

[1] Rüdiger Safranski, 2014, *Romanticism: A German affair*, trans. Robert E. Goodwin, Evanston: Northwestern University Press, 14.

K. C. Snow, *Our Inevitable Third Spinoza Controversy*,
https://doi.org/10.1007/978-3-032-03045-0_1

was still nascent, the interesting and fertile period in German thought of the last years of the eighteenth century, and more specifically, to a philosophical debate called the Spinoza Controversy.

The Spinoza Controversy in some ways typified the rapid change (and reactions against change), sense of crisis yet also excitement, and extraordinary intellectual productivity characterizing its era. The Controversy also featured an important radical disjuncture between those who opposed, and those who supported (and were helping to create), a new type of Western naturalism emerging as the Enlightenment waned. In this book I will endeavor to show the unexpected richness that rethinking the Controversy can bring to environmental philosophy, what the Controversy implies about the possible future direction of Western theories of nature as a single whole, and the way the Controversy's terms crystallize the stark environmental either/or before us in the twenty-first century.

Philosophically speaking, the years from 1781 to 1794, which formed the decade-plus leading up to the emergence of early German romanticism, saw German thinkers responding to two dominant philosophers: Kant, whose first *Critique* was published in 1781, and Spinoza, whose school of thought was re-emerging into the mainstream, if in reinvented and distorted form, after many decades of quietly building subterranean influence.[2,3] As Beiser notes, in a sense Kant and Spinoza were seen, during this time, as offering the two main alternative ways of addressing the fact that the "authority of reason" to continue to supplant revelation as a fulcrum upon which to newly and rationally "justify morality, religion, and the state" and "at least in theory, explain everything in nature" had been challenged and unsettled by Kant's and others' blows against traditional Enlightenment thought.[4] This was the backdrop against which the Spinoza Controversy broke out in 1785 and continued for many years.

The Spinoza Controversy essentially pitted those skeptical of the new strand of Spinozistic naturalism, with its claims to offer a complete account of being, against those who in varying ways (and to varying degrees) supported or sought to develop and advance it. On the skeptics' side stood

[2] Frederick Beiser, 2002, *German idealism: The struggle against subjectivism, 1781–1801,* Cambridge: Harvard University Press, 1: 48, 50f.

[3] George di Giovanni, 2016, Introduction: The unfinished philosophy of Friedrich Heinrich Jacobi, in Friedrich Heinrich Jacobi, 2016, *The Main Philosophical Writings and the Novel* Allwill, trans. G. di Giovanni, Montreal & Kingston: McGill-Queen's University Press, 8.

[4] Beiser, *German Idealism,* 1.

most importantly the German philosopher and essayist Friedrich Jacobi; those he took on in the Controversy were diverse, spanning a range of positions from the aging Enlightenment mainstay Moses Mendelssohn who tolerated only some diluted elements of Spinozism, but defended the overall utility of causal reasoning and logical deduction in knowing nature and God, to those who more actively espoused Spinozism and hoped to make it remain continuous with, and yet improve upon, the dreams of Enlightenment reason while accepting Kant's critiques – and even some of Jacobi's, once the Controversy got underway. The latter cadre notably included the early German romantics who developed the outlines of a new philosophy of nature, including most notably Schelling (in that period of his work), Friedrich Schlegel, Hölderlin, and von Hardenberg, usually known as Novalis.

Technically the Spinoza Controversy took place across three distinct public debates during the period from 1785 to 1812. The first episode was the most famous and influential over a generation (or more) of German thinkers, including these romantics; all three installments were fairly consistent. I use "Spinoza Controversy" in this work to refer to all three distinct episodes during the period 1785–1812 as a single thematic debate; some other writers use "Pantheism Controversy" as the umbrella term.[5] "Spinoza Controversy" appears more accurate, because it is thinking like a Spinoza (or at least like a neo-Spinoza), i.e., in this context, trying to create and apply a rationally-derived ontological system or set of claims

[5] At times, again, the term "Pantheism Controversy" is used interchangeably in the scholarship with "Spinoza Controversy", and/or used to refer to the 1785–1789 episode; for these alternate terms cf., e.g., Julia A. Lamm, 2010, Romanticism and pantheism, in *The Blackwell companion to nineteenth-century theology,* Ed. D. Fergusson, Chichester: John Wiley & Sons, 172f; or Shane Weller, 2011, *Modernism and Nihilism,* Basingstoke: Palgrave Macmillan, 20f.

Burckhardt calls all 3 controversies the Spinoza Controversy or Spinoza Quarrel rather than the Pantheism Controversy, cf. Frederick H. Burckhardt, 1940, Introduction, in Johann Gottfried Herder, 1940, *God, Some Conversations, trans.* F. H. Burkhardt, Indianapolis: Bobbs-Merrill, 33.

Di Giovanni calls the controversies the Spinozism Controversy, the Atheism Controversy, and the Pantheism Controversy, cf. Jacobi, *Main Writings,* 630, note no. 27.

Livieri calls the three episodes the "Pantheism Controversy (*Pantheismusstreit,* 1785), the Atheism Controversy (*Atheismusstreit,* 1798–1799), and the Theism Controversy (*Theismusstreit,* 1799/1811–12)", cf. Paolo Livieri, 2024, A Study of F. H. Jacobi's 'On the Divine Things and their Revelation,' in F. H. Jacobi, 2024, *On the Divine Things and their Revelation,* trans. P. Livieri, Montreal & Kingston: McGill-Queen's University Press, 1.

regarding [all of] what really is, that constitutes the vital issue at hand in the Controversy's debates. An alternative, if we were to choose a thematic title, might rather be to call it the "Nihilism Controversy", as it perhaps might be known if Jacobi had carried the day instead of the new Spinozists. All this being said, I will briefly sketch the three episodes here as a series, for historical clarity, before returning to using "Spinoza Controversy" to refer to all three debates as one. It should be noted this sketch of the Controversy is by no means exhaustive, and also focuses most on tracing Jacobi's and the German romantics' positions. It is Jacobi's arguments and their continued relevance that this book is focused on throughout, aiming to draw out and expand upon Jacobi's positions particularly as they form a potential foundation for a new type of philosophy of nature (or non-philosophy of non-nature, as Jacobi might have it), one which can, among other things, finally cast a different light on why the West, as of the early twenty-first century, is so able to continue to slide into an environmental void of its own making with so little capacity to change course.

The Spinoza (or Pantheism) Controversy of 1785–1789

The Spinoza Controversy began in 1785, when Jacobi (b. 1743–d. 1819) and Mendelssohn (b. 1729–d. 1786) published warring accounts of whether their mutual friend, the playwright G. E. Lessing (b. 1729–d. 1781), had been a Spinozist, and if so, what exactly this entailed. Jacobi's position was that if Lessing *had* been one, as Jacobi inferred to be the case from some of their conversations, then Lessing had deeply erred, since all monist, logically-derived systems were absurd if held to be true and complete accounts of the really existing world. Mendelssohn's position was that he did not believe Lessing had been a covert follower of Spinoza, but that if he had been, it would be possible to resolve much of the basic approach and findings of Spinoza with Lessing's theism and Enlightenment faith in reason, and indeed Mendelssohn believed Lessing may have been actually working out this very concordance as a kind of pet theological project, albeit one he had never completed.

Jacobi's first publication on the Controversy was "Concerning the Doctrine of Spinoza in Letters to Herr Moses Mendelssohn", usually referred to as the Spinoza Letters or *Spinozabriefe*; Mendelssohn's opening salvo was contained in "Morning Hours of Lectures on the Existence of God". In 1786, Mendelssohn replied further to Jacobi with "To

Lessing's Friends", and Jacobi published "Reply to Mendelssohn's Accusations concerning the Letters on the Doctrine of Spinoza[6]".

In 1787 the theological writer J. G. Herder waded in with *God, Some Conversations,* in which he presented his own view of Spinoza's philosophy with the stated aim of rehabilitating Spinoza from Jacobi's charges of atheism and re-situating Spinoza within the new post-Newtonian physics based less on "corporeal matter" and more on energy and "active or interactive forces[7]". The newer physics in Herder's view could help "correct" Spinoza's use of what for Herder were outdated Cartesian-influenced categories of thought and extension,[8] categories partly responsible for Spinoza's mis-construal as a materialist.[9] Herder's reading of Spinoza rendered Spinoza's God the source of "an infinite number of forces" behind both thought and matter.[10] He tried to define Spinozan necessity as not blind, but as an expression of this higher type of intelligence, and wrote that a true picture of Spinoza's divine "cosmic system" must elicit an "admiration [which] is rational", replacing "devout but empty wonder[11]".

In 1789, Jacobi published a second edition of the *Spinozabriefe*, in which he added some more replies to Mendelssohn and some direct critiques of Herder in response to *God, Some Conversations*, which Jacobi characterized as more of the increasingly popular "poetical philosophy which likes to waver midway between theism and Spinozism[12]". Jacobi faults Herder's understanding of Spinoza *simpliciter*, too, claiming that Herder distorts Spinoza's positions.[13] Jacobi also included excerpts from *On the Cause, the Principle, and the One* by Giordano Bruno, intending him as another object lesson in the difficulties arising in any monist rationalized naturalism. Another new supplement in the 1789 edition of the

[6] Cf., e.g., David Bell, 1984, *Spinoza in Germany from 1670 to the Age of Goethe,* London: University of London Institute of Germany Studies, e.g., 71.

[7] Herder, *God,* 102–103 (Conv. II, no. 450); 91–92 (Conv II, nos. 432–433); 95 (Conv I); see also Conv. II, no. 438.

[8] Herder, *God,* 101f. (Conv. II, nos. 447f.).

[9] Herder, *God,* 226 (Re: Conv. II, no. 457 (1)).

[10] Herder, *God,* 103 (Conv. II, nos. 449–451).

[11] Herder, *God,* 131–132 (Conv. III, no. 490).

[12] F. H. Jacobi, 2016, Concerning the Doctrine of Spinoza in Letters to Herr Moses Mendelssohn, in Jacobi, *Main Writings,* 364, 338–339.

[13] Cf., e.g., Jacobi, Concerning the Doctrine, 365–366. Bell points out that Jacobi presents Spinoza as consistent – too consistent to need, for example, Herder's interpretation or updating; cf. Bell, *Spinoza in Germany,* 81–82.

Spinozabriefe was a letter written by the Dutch philosopher Hemsterhuis (b. 1721–d. 1790).[14]

Hemsterhuis and the German philosopher J. G. Hamann (b. 1730–d. 1788) were two of the thinkers with whom Jacobi associated his views, though they refrained from clearly backing Jacobi. Hemsterhuis in 1779 had opposed resting any idea of divinity on mathematical relationships which are obvious and certain to humans only through being so circumscribed and finite, noting that "those who wish to prove the Divinity by way of the small amount of the so-called order they perceive build … on shaky ground … it seems to me that God and Order must be proven in completely different ways [from how this so-called order is proven][15]". The universe's Order "is quite different from what we call order"; the so-called order detected by our mathematical, logical type of reason cannot be used to prove either God's existence or represent this other, ultimate, Order; while humans "see their order only in everything that is close to them, on the surface of the earth or in the motions of the planets around their sun," in the "vastness of the starry vault … [there is] perfect disorder[16]".

Hemsterhuis further foreshadows some of Jacobi's positions when he notes that "*Necessary* is a word that philosophers have strangely abused" since "the word *necessary* is only an epithet added to what is"; "to be, to act, to produce, to persist necessarily, says nothing other than to be, to act, to produce, or to persist[17]". In *Letter on Fatalism*, Hemsterhuis writes that the notion of building a complete explanation of the world using necessary causes leads to an "absurdity", since depictions of causes necessarily leading to effects can only be accurate about the past, and not the future, since "the future cannot exist … as it is still future;" this breaks apart the system of a necessary universe.[18] Elsewhere he faults the "abuse of

[14] A Dutch philosopher and writer on feeling and aesthetics who resisted formal philosophy, Hemsterhuis influenced Jacobi, Herder, Novalis, and Schlegel. Cf., e.g., H. Moenkemeyer, 1977, François Hemsterhuis. Admirers, critics, scholars, *Deutsche Vierteljahrsschrift Für Literaturwissenschaft Und Geistesgeschichte,* 51: 505.

[15] François Hemsterhuis, 2022, Aristaeus, or on the Divinity, in *The Dialogues of Francois Hemsterhuis, 1778–1787*, Volume 2 of *The Edinburgh Edition of the Philosophical Works of François Hemsterhuis*, eds. Jacob van Sluis and Daniel Whistler, Edinburgh: Edinburgh University Press, 66.

[16] Hemsterhuis, Aristaeus, 67.

[17] Hemsterhuis, Aristaeus, 72.

[18] F. Hemsterhuis, 2023, Letter on Fatalism, in *The Philosophical Correspondence and Unpublished Writings of François Hemsterhuis*, Volume 3 of *The Edinburgh Edition of the Philosophical Works of François Hemsterhuis*, eds. Jacob van Sluis and Daniel Whistler, Edinburgh: Edinburgh University Press, 72–73.

abstraction" and the "absolute and despotic empire by which the genius of Plato and Aristotle subjugated the entire world," with Socrates the author of the "only good and true philosophy[19]".

In Hemsterhuis' *Letter on Atheism,* which Jacobi included in the second edition of the *Spinozabriefe,* and which was written at Jacobi's request but addressed to their mutual friend, Princess Gallitzin,[20] Hemsterhuis notes that humans' limited observations of matter have led them astray; "the richness of what [man] calls matter in any category is only analogous to the small number of his current organs, and … this matter, in which his imagination believes it can see the Whole, is only an infinitely small [part] of the Universe[21]". Europe for Hemsterhuis is in the age of an atheism "born from the vanity of the triumphant intellect" which tries to use mechanics and Newtonian laws to go beyond the limits which Newton himself acknowledged inhered in his own discoveries.[22] Because humans know so much, relatively speaking, about the "matter" they are able to define, and so little about so much else they encounter, they make matter the substance of their claimed "world" and forget that this resulting world is not the actual, real one (which it could only be if the world really and in fact were constituted solely of such "matter"). They ignore the fact that "*matter* is but a word which designates all real essences as they relate to our current organs", a word which functions really only as "a sign for essences as known[23]".

Hamann also provided fodder for Jacobi's opposition to naturalism, and Jacobi's emphasis instead on what we might call, adapting and slightly changing Sandkaulen's term, "metarational cognition[24]". Sandkaulen claims that the "younger generation of … post-Kantians," idealists like Fichte and Hegel, were influenced by Jacobi in that they developed "a

[19] Hemsterhuis, 2023, Letters on the History of Philosophy up to Spinoza, in *Philosophical Correspondence,* 78–79.

[20] Cf. di Giovanni in Jacobi, *Main Writings,* translator's note, 360–361. Again, Hemsterhuis appeared loathe to become directly involved in the fray, preferring to keep the Princess as his stated interlocutor.

[21] Hemsterhuis, 2023, Letter on Atheism from Diocles to Diotima, in *Philosophical Correspondence,* 108.

[22] Hemsterhuis, On Atheism, 107–109.

[23] Hemsterhuis, On Atheism, 109.

[24] Birgit Sandkaulen, 2023, Jacobi and Spinoza, trans. Joseph Carew, in *Friedrich Heinrich Jacobi and the Ends of the Enlightenment: Religion, Philosophy, and Reason At the Crux of Modernity,* ed. Alexander J. B. Hampton, Cambridge: Cambridge University Press, 35.

new understanding of metarational cognition modeled on Jacobi[25]". I would argue (and will endeavor to make my reasons why clear in what follows) that while Jacobian metarational cognition is not actually amenable to the kind of co-option to which it is subjected by his peers and successors, we also find an attempt to co-opt (or learn from) Jacobi's preoccupation with metarational cognition in the early German romantics. Regardless of how other philosophers reacted to or reshaped the concept, what I will offer in what follows is that what Jacobi, along with Hemsterhuis and Hamann, is most preoccupied with in terms of metarational cognition is preserving an independent, free, and person-seated faculty of reason which is not bound to the theoretical understanding's identitarian, mathematical-definitional type of truths or the ways we have of perceiving these [as rationally "true" in their own sense]. Metarational for Jacobi still means reasoning, as we will see; the metarational could be said to be "meta-" (or beyond, outside, and able to reference or comment upon, per this prefix's dictionary definition) only the logical and deductive *sub*-type of reason which Jacobi calls the theoretical understanding. For Jacobi, the metarational or metatheoretical faculty is crucial for the subject because it is what allows the subject to experience and encounter the world as truly external, without the immediate imposition of theoretical or deductive certainties made by the human mind, including those turning on some type of perceived necessity. In effect it is for Jacobi the only part of reason which can remain anchored in the real world that is, and not risk becoming lost in its own ideal constructions. In 1784 Hamann defines the "apodeictic," i.e., absolute, "certainty of mathematics" as dependent "mainly on a curiological,[26] so to speak, portrayal of the simplest, most sensible intuition and then on the ease of proving and representing its synthesis and the possibility of its synthesis in obvious constructions or symbolic formulas and equations[27]". Mathematics' "universal and necessary reliability" in itself is no more remarkable than the "infallible and unerring instinct of insects[28]", yet the fact of this certainty is overinterpreted, including as proof of a connection between mathematics and some hypothetical realm of essences more real than what really exists.

[25] Sandkaulen, Jacobi and Spinoza, 34–35.

[26] "Curiological" is defined by Merriam-Webster as "representing things by their pictures instead of by symbols – used of hieroglyphic writing."

[27] Johann Georg Hamann, 2007, Metacritique on the Purism of Reason (written in 1784), in *Hamann: Writings on Philosophy and Language*, Ed. K. Haynes, Cambridge: Cambridge University Press, 209–210.

[28] Hamann, Metacritique, 210–211.

Mendelssohn's Positions

Mendelssohn's *Morning Hours* lectures contain an Enlightenment defense of "speculation" and the utility of logical reason in pursuing "knowledge of actual things[29]". He discusses the existence of necessary versus contingent beings,[30] continuing his quasi-Leibnizian, deistic, support of the distinction between the two, and the further distinction between logical necessity and moral necessity, where the first is a "blind" force but the second actually constitutes a type of freedom[31]. He denies that some moral ill or logical error must result from being a rationalizing world-builder in the same way as Spinoza. His is a kind of high Enlightenment, pre-critical approach, not daunted by using "exist" as a logically manipulable and analyzable predicate, beginning with serving as a real predicate for the thinking subject.[32]

In a lecture on Spinozism, Mendelssohn agrees with what he calls Spinoza's representation of "contingent beings as ... beings that by their nature presuppose for their existence a regressive series without beginning and that in the forward direction endlessly advance the series into actuality," but demurs from Spinoza's monism which would do away with contingent things altogether; "contingent things ... we say, have their own substantiality outside of God, although they can only exist as effects of His omnipotence[33]". In this and further cases, e.g., regarding the differences in infinite extension and infinite power,[34] Mendelssohn may differentiate himself from Spinoza, but he is still willing to engage with him directly, and does not outright reject the notion of a rationalized world-explanation as such. This marks his key distinction from Jacobi, whose main arguments, as we will see, turn on trying to show that Spinozist ontologizing is still effectively an internalist exercise that cannot make contact with the really existing world—even while simultaneously granting Spinoza's original system (which Jacobi thinks, with some justification, that few of his opponents actually fully understand or have even studied directly themselves) a kind of special status as the most nearly perfect philosophy of its type.

[29] Moses Mendelssohn, 2012, Morning Hours, or, Lectures on the Existence of God [1785], in *Last Works,* trans. Bruce Rosenstock, Champaign: University of Illinois Press, 16.

[30] Mendelssohn, Morning Hours, 84ff. (Lec. XII).

[31] Mendelssohn, Morning Hours, 87 (Lec. XII).

[32] Mendelssohn, Morning Hours, 87–88 (Lec. XII).

[33] Mendelssohn, Morning Hours, 93–94 (Lec. XIII).

[34] Mendelssohn, Morning Hours, 97f (Lec. XIII).

Mendelssohn considers there to be a form of "purified pantheism" he might be able to ascribe to his late mentor, Lessing, without any great scandal resulting; if Lessing were exploring any form of pantheism, he surmises, it would be "a completely refined pantheism … in the finest harmony with everything that can have an impact upon our lives and our happiness[35]". He believes Lessing may have actively been engaged with the task of working on further reconciling this "purified" type of Spinozism or pantheism with Lessing's own theism, noting that, "indeed … he was on his way to connecting pantheistic concepts even with those of positive religion, and this synthesis was every bit as successful as was the emanation of the ancients, which for many centuries had been incorporated into religion as the only orthodox doctrine[36]".

In *To the Friends of Lessing*, Mendelssohn defends Lessing as a "great and admired defender of theism and of the religion of Reason" against the "indictment" of Jacobi[37] which clearly upset him to the core, and according to some accounts hastened Mendelssohn's own death.[38] Mendelssohn once more presents ways in which Lessing was most likely misunderstood or misrepresented by Jacobi: either Jacobi did not understand the "purified Spinozism" that Mendelssohn believed would be the most that could rightly be attributed to his friend Lessing; or Lessing might have studied or followed Spinoza merely as a side-interest, not as a source of core beliefs; or, again, Lessing may have been carrying out additional, as-yet unpublished or incomplete, theological work linking Spinozism and pantheism with Christian belief, the ultimately Christian aims of which must absolve Lessing of any taint of atheism or heresy.[39] Mendelssohn also hazards that Jacobi's reported exchanges with Lessing may have been a case where the latter was merely conducting a hypothetical argument with Jacobi for intellectual curiosity's or politeness's sake. Mendelssohn redoubles his commitment to his own speculative "task of correcting the claims of common sense and, as much as possible, turning them into rational knowledge," noting that "as long as both of them, speculation and common sense, remain on good terms, I gladly follow them wherever they lead me[40]". Mendelssohn also takes issue with some further details on

[35] Mendelssohn, Morning Hours, 119 (Lec. XV).
[36] Mendelssohn, Morning Hours, 119 (Lec. XV).
[37] Moses Mendelssohn, 2012, To the Friends of Lessing, in *Last Works*, 148, 149.
[38] Cf., e.g., Engel, 2012, Preface, in Friends of Lessing, 141–145.
[39] Mendelssohn, Friends of Lessing, 149–150.
[40] Mendelssohn, Friends of Lessing, 158.

which Jacobi faults Spinoza, again not engaging with Jacobi's main point of contention, which is that no rational ontology could reach the real, because of the sub-type of reason upon which any systematic and logically derived ontology must rely, and this type of theoretical-speculative reason's considerable and inescapable limitations.

Jacobi's Main Views

Jacobi (again, unlike many of his opponents in the Controversy) had studied Spinoza's own work in depth, and he admired it greatly as an exhibition of a certain type of rational thought.[41] Jacobi carried out such a diligent opposite of a straw-man argument against the Dutch thinker that many who followed and participated in the Controversy believed Jacobi to actually be a Spinozist himself[42]. The focus of Jacobi's animus was essentially *any* philosophical, rational, "complete" ontology of nature or the world[43] derived from the application of theoretical (logical) reason. While Jacobi represented Spinoza's positions more faithfully than most of those claiming to feel only admiration for Lessing's purportedly chosen philosophical system, Jacobi was adamant that even Spinoza's own, rigorous, original conception of the universe could not be held to be really and objectively true of the world, no matter how internally well-constructed it was. The fulcrum of the Controversy was not that Jacobi objected to parts of Spinoza's system, but rather that he found radically, existentially upending, and simply incorrect, any belief like Lessing [may have] held that Spinoza's ontology, or any rationalized ontology, could be really somehow true of the actual world around us and somehow successfully explain this world—explain it as really existent, and as the locus of all real being including our own as human persons. Jacobi's main preoccupation was with adequately conveying the sheer "inability of philosophical reflection to

[41] Cf. also on this point Sandkaulen, Jacobi and Spinoza, 37–42.

[42] Cf. Sandkaulen, Jacobi and Spinoza, 42; or di Giovanni, 2016, Introduction, in Jacobi, *Main Writings,* 82; or Benjamin Lazier, 2008, *God Interrupted: Heresy and the European Imagination between the World Wars,* Princeton: Princeton University Press, 94.

[43] In this context we can define an *ontology* of nature as an explanatory account seeking to canvas or describe all that really exists or is, along with at least an implicit (and often an explicit) causal necessity-based explanation of how it came to be and what it means for it to be. I use the intransitive verb, to ontologize, throughout this work as well, since it was often very specifically the act of the human subject creating these accounts, and the various steps it required the thinking "I" to go through, which Jacobi critiqued.

reach out to existence", in di Giovanni's phrase.[44] His position was to try to show that this inability was itself something reason could grasp, and know as true; but only reason in the fuller sense, metarational (or metatheoretical) reason unrestricted to consisting solely of what Jacobi called the [theoretical] understanding, which used relations of logical necessity and mathematical certainty alone to derive truth claims. Jacobi was trying to identify a problem, a tendency toward misuse, concerning our theoretical understanding, but it was difficult for him to explicate, because it was a problem which did not emerge unless one took a perspective outside the demands of theory itself. This seems to have been part of why the Controversy was so protracted and full of misunderstanding. Most of his contemporaries, even those sympathetic to his positions, seemingly remained somewhat aloof from Jacobi's efforts to scrutinize theoretical reason as such from the imagined perspective of the remainder of reason and of the person. Even much twentieth- and twenty-first-century Jacobi-directed scholarship tends to dismiss this division or at least insist on treating it analytically, thereby effectively dismissing it just the same. Some Jacobi scholars fault Jacobi, for example, for using a Spinozan concept, intellectual intuition, which only has meaning within the Spinozan system;[45] or claim that Spinoza's philosophy overlaps substantially with Jacobi's positions[46]; or otherwise drift toward altogether ignoring Jacobi's claim to approach Spinozist naturalism truly perpendicularly, at most giving this claim *en passant*, descriptive acknowledgment. This work will endeavor not to repeat this tendency. And while Plato, as well as contemporaries of Jacobi's such as (per above) Hamann and Hemsterhuis, could be said to be read by Jacobi with enthusiasm, these thinkers cannot be made Jacobi's proxies in the Controversy context. Jacobi should be read, I will try to offer here, more singly, if we are to grasp the radical quality of his opposition to how [even ostensibly Spinoza-based] naturalism leaves the subject marooned at such an extraordinary distance from the really existing world. (We should note that in this work, we will refer to the "subject" in the Jacobian sense of the open-ended, really living, person or full self, the seat of complete subjective individuality including all the forms of reason, intuition, perception, memory, sense of beauty, and so forth, it may possess – not in the later more normatively negative and overly restrictive postmodern sense.)

[44] Di Giovanni, Introduction, in Jacobi, *Main Writings*, 9.
[45] Di Giovanni, Introduction, in Jacobi, *Main Writings*, 85.
[46] Sandkaulen, Jacobi and Spinoza, 39 f.

In the *Spinozabriefe* Jacobi discusses the fatalism and materialism which results from removing final causes in Spinoza's approach; when we then try to apply this lens to our real experience we are reduced – even when in a state such as appreciation of beauty – to believing we are merely "witnessing" "blindly self-developing forces" at work in the world and in ourselves; true individual emotion and free will must be understood as only an "illusion".[47] But this is only the entrée to his real argument, and those like Mendelssohn and Herder (and the early German romantics) who sought to, in response, determinedly, and even passionately, re-cast these blind forces as something ennobling of the person to consider as the ligaments of nature, seem to have mistaken which dissonance most concerns him. It is not that the blind forces counter and cancel or otherwise restrict the emotions and the will, but that blind forces are a different type of entity, accessed and understood differently, from emotion and will; theoretically derived entities do not connect with reality the same way immediate experiential facts and feelings incontestably do. It summons an entirely different type of thing when we discuss a blind force, compared with discussing a person (or, for Jacobi, a person-like divinity) or anything else incontrovertibly real. Jacobi holds the notion of a real world that is really only blindly self-developing forces to be incoherent for this reason, not because he wants to defend the power of the will and emotions to shape human behavior against the power of blind forces to do the same in some type of choice or struggle between two like kinds. We know the will and emotions really do make up our experience of the world outside us; we cannot subjugate them to a mere conceptual scheme which is not of the same cloth, regardless of any interpretation we could (or could not) then attach to the scheme to make it more morally or aesthetically palatable. The world "made" of only logically specifiable, provable necessity-relations is specious not species; it is a false outcome, something that results from the "thinking I" misunderstanding the tasks for which the merely theoretical faculty is suited, and ignoring its unavoidable limits. This is not a mistake, again, that the theoretical understanding itself can judge or even see, but rather a failure of the subject's fuller reason, an error made by the whole thinking person or self. To think that we can actually world-build using mathematical or logical relations – relations always defined via specifying that something *must* follow *of necessity* from something else – and have the result somehow actually "be" the real world of experience, the world that

[47] Jacobi, Concerning the Doctrine, 189.

is, is a kind of committing of a category mistake that only this fuller reasoning power can make (or detect when others make it). This mistake cannot be rectified or answered through any tweaks to (or aesthetic glosses placed upon) the concepts that sit *within* the world-"system" of blind forces arranged according to our lower reason's notions of necessity. For Jacobi, our fuller reason and our theoretical reason must always diverge at the threshold of system-building: only the latter would seek to both systematize through necessity relations and then also assert the result as being the same thing as the really present world. The fuller part of our reason, our reason which is meta-rational (where, again, "rational" denotes here only the theoretic, deductive, closed-loop certainty-constructing capacity), is what can seek to turn the merely rational faculty back from trying to systematize and ontologize, or, if the theoretical faculty insists on proceeding to "world"-build, levy its criticism on the end result. Those who refuse to read Jacobi on his own terms are perhaps usually pre-committed to the very conflation of reason as a whole with logical necessity-based argument that Jacobi was trying to disentangle. This is perhaps his fate as the "losing" side in the Controversy. The arrival of our own twenty-first century iteration of the Spinoza Controversy offers a fresh opportunity to not only reconsider his positions but also to at a minimum read him more fully within his own frame, in so doing adding to our understanding of the fissure in Western thought the Controversy represents.

Returning to his positions as expressed in the original salvos of the Controversy: this initial main argument about the non-ontological quality of human-designated relations of necessity leads into another claim, equally if not more important for our purposes in the current work. World-building based on stating logical inter-relations, when insisted upon as a really existing world (and as the only typology for anything to be really existent), is not just a passive type of category mistake, but entails an act of *nihilism* – Jacobi being among the first to use this term in something like its contemporary sense.[48] This is because for Jacobi, in order to create a "world of reason" we must always first:

> appropriate the universe by tearing it apart, and creating a world of *pictures*, ideas, and words, which is proportionate to our powers, but quite unlike the real one. We understand perfectly what we thus create, to the extent that it is our creation … [but] whatever does not allow being created in this way, we do not understand.[49]

[48] Cf., e.g., Lamm, Romanticism, 172 f.; Weller, *Modernism*, 20 f.

[49] Jacobi, Concerning the Doctrine, 370.

Even before this "tearing it apart" can begin, the thinking "I" must also first pull itself away from the real world around it which is characterized not by an "I" but by our intuition of the "Thou": our relations with both God and the external world (and other people), whose reality we can know "with the very same certainty with which we become aware of ourselves, for without the Thou, the *I* is impossible[50]". For Jacobi this pulling away is always our first major error, setting up the thinking I to convince itself it can be a world-builder at least in principle, when we are actually only ever able to connect with the really existing world as a receiver. Elsewhere he speaks of the beginning phase of theoretical world-building as requiring the human to seize "his life by the root so to speak, in order to throw it from him[51]", before he can start to world-build using causal necessities and their interlinkages alone. The insistence that we can include the full existence of the real world, including its being as such, in any conceptual schema this falsely isolated "I" patches together out of its comprehended terms entails the actually existing world's annihilation,[52] including the annihilation of the type of communion we had with all the "Thous" of our pre-conceptual acquaintance. From the metarational perspective, only irrationality and unreason can follow, no matter how order-filled, "necessary," "correct," or "true" the specious world itself may be when judged by its own internal standards and rules.

Jacobi envisions the scientist's job being to "unveil *existence*, and to reveal it," but cautions, somewhat in the mode of Hume, that the scientist cannot take his "obsession with explanation" beyond where it can rightfully claim to reach; he must not let himself be blinded by the false sense of absolute or complete explanation of all things which can come from his "*joining* and *hanging* together only what is explainable in things[53]".

Jacobi's key goal was to hold up, in contrast to the new Spinozistic naturalistic philosophy, his own position he later, in 1799, christened his "non-philosophy" of "*non*-knowledge[54]", but which is also already evidently his position here: what we require at the center of our knowing is

[50] Jacobi, Concerning the Doctrine, 231.

[51] This is from numbered paragraph no. 316 of a Jacobi text not strictly speaking usually included in the Controversy texts, F. H. Jacobi, 2016, Edward Allwill's Collection of Letters, in Jacobi, *Main Writings*, 495.

[52] Jacobi, Concerning the Doctrine, 372f., including Jacobi's footnote to no. 419, 373–374.

[53] Jacobi, Concerning the Doctrine, 194–195.

[54] Cf. Jacobi, 2016, Letter to Fichte, in *Main Writings*, e.g., 501.

the awareness of how little philosophy (in the rational, natural philosophy system-building, speculative-theoretical sense) can say about what really is. The access to the really real which is claimed (both tacitly and actively) by a philosophy seeking to rationally directly speculate about being is only a kind of illusion or even self-delusion. From the metarational perspective, any theoretically-constructed "world" is just some [deceptive] variation on saying simply: "whatever is, is[55]", because it is impossible for such a world to have real content. The "complex that we call *nature*[56]" is only defined through various conditions we invent for it and then place upon it; while we may be able to erect these and take them away again, this has naught to do with uncovering or proving anything about actual being as such[57]; the "complex" so defined can never touch upon or encompass the real. We may construct, remove, and then re-add various components according to a process of making and then testing truth-claims through logic and proof in the mold of mathematics; this, again, has naught to do with real world-making or world-shaping. For Jacobi, supernatural causes are all that can explain the world's existence. But I would argue that he first and foremost means super- to—above or outside—this type of constructed pseudo-nature, this man-created conceptual "complex." We could potentially encounter the supernatural within the really existing non-human world, as long as we have moved back into the realm of direct experience, and away from trying to see the external real world around us as really the same thing (and more importantly as really the same *type* of thing) as our artificial concept-complex.

While especially Spinozan (or Spinozistic) systems attempt to erect these complexes using necessity as a main animating and organizing principle, moreover, theoretical "reason has no concept of any necessity except the one that it establishes itself by means of its progressive and unrelenting process of separating and reuniting, by alternately retaining and letting go and finally displaying this necessity in identical propositions[58]"; necessity in the theoretical framework is just identity; it is always based on definitional and self-contained truth-claims. Any of our self-created causal complexes "made" from necessity-relations visible to and traceable by logical

[55] Jacobi, Concerning the Doctrine, 372f. (nos. 416–422), especially 373–374 (Jacobi's footnote to no. 419).

[56] Jacobi, Concerning the Doctrine, 376 (no. 425).

[57] Jacobi, Concerning the Doctrine, 376 (nos. 424–430).

[58] Jacobi, Concerning the Doctrine, 374 (no. 421).

reason – no matter if ostensibly realist, idealist, Fichteian, Spinozan, or of any other type – are identical in that when they forget that, in Förster's phrase, per Jacobi it is inherently impossible to attempt "a universal philosophical explanation or foundation",[59] and try to use their form of "necessity" as though it really existed, and existed as that which actually caused or made the really existent world, they become vectors of nihilism against the real world that is, and is apart from us, and is without need of our factitious would-be foundationalizing. The "worlds" we fully know become refusals to point and say, *it* is, refusals to acknowledge the prior-ness of the real world to reason, its radical externality to our thinking "I." Systems we construct and then endow with belief that these are the world, must then always become destroyers, because concealers and deniers, of the truth of the real world's independence from, and externality to, theory. The rawly independent world, in which we are all originally "rooted", in his metaphor, is only perceptible, for Jacobi, via our metarational, fuller reason, the part of the self which has to decide, first and fatefully, whether to seize "his life by the root so to speak, in order to throw it from him[60]", in order to then attempt [futilely] to reconstruct being using the tools of logical necessity. This reconstruction process, once indulged in, becomes boundary-less and self-perpetuating, because the theoretical faculty is in itself unable to recognize its unlikeness with reality, and will seek to encompass the vast natural universe as a whole in what it claims to explain. For Jacobi, as we will return to in Chap. 2 (cf. Chap. 2, last section), this entails in effect cognitively draping grotesque half-sketched *nihils* of pure reason like screens around our awareness, hyper-extending the claimed descriptive or ontological success of what are in fact extraordinarily sparse accounts. This placement of a nihil of reason, a nothing or a non-explanation masquerading as a something and as positive knowledge, where the infinite universe really is, is an act of nihilism or annihilation which is complementary to, and completing of, the nihilistic act with which world-building using theory must always begin.

Intertwined with arguing that even Spinoza's world-construction cannot escape being nihilistic because it is real-world-replacing in this way, Jacobi does, as we have noted previously, spend time holding Spinoza's philosophical effort aloft, painting it as the best systematic approach to being as such he has read, one which exceeds shallower systems (cf. also

[59] Eckart Förster, 2012, Goethe's Spinozism, in *Spinoza and German Idealism*, eds. Eckart Förster and Yitzhak Y. Melamed, Cambridge: Cambridge University Press, 86.

[60] Jacobi, Edward Allwill's, 495.

Sandkaulen on this point[61]). He rejects what he considers Spinozist naturalism,[62] again, not because of one or more problems he finds within Spinoza's own system, or because of Jacobi's contemporaries' failure to understand Spinoza's thought, but because Jacobi's appreciation of Spinoza's rigor and system-building skill (seen within its own frame of reference) convinces him that no thinking of this type could ever break out of the limits inherent in it. For Jacobi, even Spinoza had erred in his *belief* that the theoretical understanding was a tool useful in fully laying out a systematic ontology of nature (including human beings) and God – of all of what really exists. Even Spinoza just ends up demonstrating something about what one sub-faculty of human reason could [and could not] do, not anything about the world itself. If we consider that Spinoza was an inheritor of medieval monotheism's quite different types of proof-like treatments of questions turning on God's existence, this is not surprising; the effort to prove that there must be a self-caused or perfectly necessary being, that it must be God, and that God must therefore be real, has a long history from that time of serving more as a theological accoutrement or demonstration of faith (and of how faith and reason could coexist) than a "proof" in the modern sense of the term. In the Enlightenment context, and particularly in Spinoza's able hands, rationalized ontologizing arguably pressed the limits of this tradition, setting up Spinoza's monist array to be read differently (and in more of a realist than a theistic vein), but then, in Jacobi's view, fell into absurdities as a result. Spinoza's system was maximally alluring due to its intellectual rigor and power[63] and its monistic seamlessness; it satisfied Jacobi as a limit-case of the type of rational philosophical full accounting of the world—the type of lofty taking-aim at the "absolute inclusion of all of reality" in a single reasoned account, to use Sandkaulen's gloss on how Jacobi read Spinoza—which Jacobi opposed.[64]

[61] Sandkaulen notes that Jacobi truly gave Spinoza his due, not only through the significant number of years he spent reading and studying him, but also, e.g., in grasping that "Unlike in [Wolffian, e.g., or simple] rationalism, Being in Spinoza … is no abstract conception, no *ens rationis,* but is rather, in its capacity as 'Absolute Being,' the absolute inclusion of all of reality", cf. Sandkaulen, Jacobi and Spinoza, 41, also citing *JWA* 2.1:44-7, *MPW* 282-5.

[62] Per, e.g., Douglas, Spinoza himself was not a "naturalist" in the contemporary usage, cf. A. Douglas, 2015, Was Spinoza A Naturalist?, *Pacific Philosophical Quarterly*, 96: 77–99. Jacobi's concern is mostly over how Spinoz*ism* is manifesting in Jacobi's own milieu, roughly a century after Spinoza's life, and how he foresees naturalism proceeding going forward, if this use of Spinoza for these new naturalistic ends is not checked. But to check this first generation of offspring, Jacobi begins by, as it were, addressing the parent.

[63] On this point cf. also Livieri, A Study, 1ff.

[64] Cf. Sandkaulen, Jacobi and Spinoza, 41 (and footnote no. 61 above); 43; 47.

The Atheism Controversy of 1798–1799

The second round of what I am here referring to as the Spinoza Controversy emerged as a swirl of negative attention directed toward the German idealist philosopher J. G. Fichte around 1798, leading to his being accused of atheism and fired from his teaching position in Jena in 1799.[65] Fichte was popular with his students and seemed to some to hint at the return of a thinking "I" able to parry Kant's questioning of the ultimate reliability and extent of its knowledge, and the public attacks on him were a prominent affair in the German world of letters. Many well-known figures, including Kant, who publicly denounced Fichte, weighed in, as did Jacobi, in a 1799 "Letter to Fichte" (or "Open Letter of 1799[66]"). As Martin describes them, the "documents pertaining to … [the] so-called Atheism-Controversy are enormously diverse, comprising official documents … formal petitions from student organizations, denunciations from anonymous pamphleteers, poignant private reflections, and a series of ill-timed and ill-considered public statements from Fichte himself.[67]"

Since our main interest in this work is Jacobi's position he developed against those who would espouse various forms of speculative (or theoretical) reason's ability to meaningfully encompass all of what is, it is not necessary to delve into Fichte's work here. Fichte's exploration of a new type of "critical idealism[68]", with its famous "I=I" proposition of identity and ranging of the thinking "I" against a world which must first be understood as the "Not-I",[69] left Jacobi nonplussed, and while as Breazeale notes, many of Fichte's contemporaries arguably misunderstood him, since his philosophical intentions were slightly more limited and more empirically aware than most gave him credit for during the Atheism

[65] Wayne Martin, 2007, Transcendental Philosophy and Atheism, *European Journal of Philosophy* 16: 109.

[66] Di Giovanni, Introduction, in Jacobi, *Main Writings,* 10.

[67] Martin, Transcendental, 109.

[68] Beiser, *German Idealism*, 349.

[69] As Safranski notes, Fichte was seeking to respond to Kant by trying to go deeper in observing the act of thinking itself; for Fichte, the "thinking I" and the "[passively] thought I" "move in an active, productive circle," such that the I is always "generating itself in reflection"; "The I is in motion, it is alive, we feel it in us." The world becomes the "Not-I" which is "given from outside … but only within the circumference of the I, beyond which it can never step, and to this extent the Not-I is itself an aspect of the I. The Not-I is a limitation, which the I takes over as a self-limitation." Safranski, *Romanticism,* 43, 44.

Controversy,[70] Jacobi's own radical opposition to any form of (even "critical") monist, rationalized, relations-of-necessity-based, philosophizing of "all that is" or "all that can be known" did not direct itself now against Fichte as a result of any subtle interpretive gradation which, if adjusted, might have led him to refrain. Fichte was for Jacobi another symbol of the type of rational ontologizing which Jacobi now continued to try to characterize as actively nihilistic. The main point at issue for Jacobi was not the details of Fichte's form of "critical idealism" per se, but simply *that* another thinker was turning toward developing a rationalized, speculative system in a spirit of believing in its applicability to real existence and in its utter self-sufficiency and completeness in being so applied. Per Martin, Jacobi used the Fichte controversy to "press his diagnosis of a trend toward nihilism in the purportedly enlightened philosophies of the day[71]".

Jacobi's arguments in the "Letter to Fichte" return to invoking, in effect, the need for the metarational faculty to make a better decision at the threshold of world-making, instead of once again, as in the earlier Spinozism case, setting theoretical reason loose to go about its deceptive processes of re-creating the existing world using logically manipulable ideas it mistakes for really "real" building blocks. *Our* world-building must proceed "by way of dissolution and analysis to *the state where nothing is outside the I*": but this act of "dissolving all *being into knowledge*[72]", is not only epistemologically unjustifiable and incoherent, but also, again, actively destroys what is by removing it from the person and sealing it away permanently from inclusion in any completed "world" according to reason. Even if, as Breazeale contends (cf. above), Fichte was merely conducting some exploratory work on exposing the foundations of viable cognition, it was still close enough to the types of speculation Jacobi considered inherently invalid to attract Jacobi's criticism. It simply is not possible, for Jacobi, to actually reduce reality to concepts of reason alone such as would be necessary to allow such a process to take place as Fichte describes;[73] notably Fichte's denial of the Thou and all the other incomprehensible aspects of reality that ground and generate rational thought. Whether Fichte in particular fully deserved his ire or not, again, Jacobi was

[70] Dan Breazeale, 2024, Johann Gottlieb Fichte, in *The Stanford Encyclopedia of Philosophy* (Fall 2024 Edition), eds. Edward N. Zalta and Uri Nodelman. https://plato.stanford.edu/archives/fall2024/entries/johann-fichte/

[71] Martin, Transcendental, 109–110.

[72] Jacobi, Letter to Fichte, 509 (no. 18).

[73] Jacobi, Letter to Fichte, 509 (no. 18).

also using his difference with the Jena professor as a device to raise his objections against a whole type of thinking that mis- and over-interpreted the products of rational speculation. He called Fichte's idealism just another version or extension of the Spinozist approach: Fichte offered merely an "inverted Spinozism".[74]

In one notable metaphor in the *Letter*, Jacobi compares Fichte's preoccupation with laying the foundations for a coherent [critical] idealist system to worrying exclusively about the origin of a stocking (something self-referentially explicable; something derivative; in this case, mere theoretical, logical statements and testable inter-relations) instead of the human leg which wears it (something inherently inexplicable by the stocking's lights, and also the entity giving the stocking its real purpose, a purpose which cannot be seen within the weave of the stocking itself).[75] Jacobi ridicules the notion that a human being, who is a full and metarational living entity, and as such can intuit and see the difference between the two (the stocking and the leg), could ever be fulfilled by Fichte's type of "love for the knowledge *of knowledge* as such, the insight into *seeing* as such, the doing of *doing* as such[76]" as would be required to prioritize knowledge of what holds together this "*mere weaving of weaving*" of the stocking,[77] and even to propose the same weaving as the stuff of reality itself. Jacobi writes: "what would [holding the world to be] a mere weaving of weaving amount to, from the animals below up to the saints above?—I declare that my reason, my whole inner self, flies into a rage before a representation of this sort, it shudders in horror and fright; I recoil from it.[78]" This is partly addressed to F. Schlegel, who had unfavorably reviewed Jacobi's philosophical novel *Woldemar* three years earlier, in 1796. In the review Schlegel had expressed appreciation for some of Jacobi's positions expressed during the earlier episodes of the Spinoza Controversy, but had faulted Jacobi for not embracing what Schlegel asserted was the new value of philosophy: the "'pure interest in knowledge and truth[79]'".

[74] Jacobi, Letter to Fichte, 502f. (nos. 3f.).

[75] Jacobi, Letter to Fichte, 509–511 (nos. 18–23).

[76] Jacobi, Letter to Fichte, 511, (nos. 23–24).

[77] Jacobi, Letter to Fichte, 511 (no. 23).

[78] Jacobi, Letter to Fichte, 511 (no. 23).

[79] Ernst Behler, 1996, Friedrich Schlegel's Theory of an Alternating Principle prior to his Arrival in Jena (6 August 1796), *Revue Internationale de Philosophie*, 50: 383–384. On 384 he quotes the F. Schlegel review of *Woldemar* as KFSA 2, 69. See also Di Giovanni in Jacobi, *Main Writings*, 625 (note no. 15).

The stocking metaphor also echoes Jacobi's concern put forward in the *Spinozabriefe* with the tendency of all philosophical systems to reduce the world to the working of "blindly self-developing forces", simply because this is how our theoretical, logical faculty pieces things together in general. For Jacobi truth must be more than simple self-consistency and self-reference, restated a number of ways. The "weaving of weaving" view only identifies various rules for "stitching", and can only ever talk about (and display) speculative reason to itself, making it clear how to follow a particular kind of rule to acquire a particular kind of overall consistency. Jacobi suggests it is the human subject's fuller [metarational] sense of reason, its "whole inner self," that is like the person—able to knit the stocking and to take it off or put it on according to when it is or is not useful or needed, but also able to concern itself with wholly other kinds of things.

Jacobi points out again that to comprehend something using the types of "necessity" accessible to rational speculation or theorizing, we must first destroy the world we wish to theorize, and then try to re-create it; but this leaves whatever we create a "*mere schema*" because it is idea, or imagining, alone.[80] Jacobi calls the only meaning a person can derive from such a schema a "naked logical enthusiasm" for it (another reference to Schlegel's *Woldemar* review); he does not know any other type of response one could have to such a stocking-world, such a pseudo-world based only upon imagining "*purely self-intending and self-contemplating activity …, without purpose, or point of arrival*[81]".

The Pantheism (or Theism) Controversy of 1807–1812

The final bout of the Spinoza Controversy came after the dawning of the new, 19th, century, when Jacobi turned his sights on Schelling as one more example of the kind of Spinozist monist naturalist and philosopher he found so objectionable. The issue, again, was not the details of the content of Schelling's thought per se, though he objected to this as well, but more its framing assumptions and the use to which it would claim to be suitably put. He saw in Schelling's philosophy (which by this point had moved out of its romantic period, though it still contained strong elements of romantic thinking) another instance of the nihilism which would

[80] Jacobi, Letter to Fichte, 508 (nos. 15–16).
[81] Jacobi, Letter to Fichte, 511 (nos. 22–23).

place the really existing world into the same category as—make it in its essence somehow the same *kind* of entity as—the outputs of the sub-faculty of reason that was our merely theoretical ability to stitch and weave together relations of logically bound necessity.

This episode began brewing after a Munich lecture given by Schelling in 1807,[82] which Jacobi perceived as, in Di Giovanni's gloss, Schelling attempting an "impossible synthesis of Platonism and Spinozism that passed off the pantheism of Spinoza under false Platonic pretences[83]". The debate broke into the public eye around 1811–1812, and is usually referred to as the "pantheism controversy" or the "theism controversy".[84] Aside from the lecture, Schelling's *Philosophical Investigations into the Essence of Human Freedom* of 1809 also spurred and influenced[85] Jacobi's initiation of the public side of the controversy, which came with his publication of *On the Divine Things and their Revelation* of 1811.[86] Besides this 1811 text, the other main text associated with this last phase of the Controversy was Schelling's response, his *Memorial for Herr Friedrich Heinrich Jacobi's Writing 'of Divine Things etc.*[87]' of 1812. Again, we will focus here on Jacobi rather than Schelling, because we are most interested in this work in understanding Jacobi's core position on the impossibility of

[82] Schelling's 1807 lecture, "Concerning the relation of the fine arts to nature", now translated as: "The Philosophy of Art, an Oration on the Relation between the Plastic Arts and Nature", was delivered after Schelling became president of the Munich-based Academy of Sciences; cf. di Giovanni, Introduction, in Jacobi, *Main Writings,* 10–11. Livieri anchors this bout of the Controversy in 1799, on account of Jacobi specifying in 1816 that his essay On the Divine Things was connected, for Jacobi, to his 1799 Letter to Fichte, cf. Livieri, A Study, 6.

[83] Di Giovanni in Jacobi, *Main Writings,* 631 (note no. 27).

[84] Cf., e.g., di Giovanni in Jacobi, *Main Writings,* 8f., 630f. Cf. also Footnote no. 5 above. Again, the pantheism qualm, for Jacobi, I would argue, had less to do with matters of Christian doctrine than the fact that in this new version of speculative pantheism (or speculative panentheism), God, who for Jacobi is the most real and least analyzable type of being, would be changed into the same type of entity as we think we "make" via manipulating and testing relations of logical necessity (and this concern would hold despite the refinement and subtlety of Spinoza's own use of this type of reasoning).

[85] Cf. di Giovanni in Jacobi, *Main Writings,* 633 (note 31).

[86] Cf. P. Livieri, 2019, *F.H. Jacobi's "On Divine Things and their Revelation": A study and translation,* ProQuest Dissertations & Theses Global, 142ff.

[87] Schelling's text is not yet locatable in English translation: in German the work is *Denkmal Der Schrift Von Den Göttlichen Dingen &c. Des Herrn Friedrich Heinrich Jacobi: Und Der Ihm In Derselben Gemachten Beschuldigung Eines Absichtlich Täuschenden, Lüge Redenden Atheismus* (Tübingen: In der J.G. Cotta'schen Buchhandlung, 1812.).

reconstructing the world, of actually ontologizing, of using logic and the theoretical, analytical faculty to reliably know and adequately represent whatever it is that has world-making [and -sustaining] power in fact, since for Jacobi, we recall, necessity and any "principle" resting upon it are only ever reflections of the criteria or conditions required by a certain sub-type of human reason in order that it can operate. Schelling, with respect to his earlier, romantic period, will reappear however in the next chapter, when we turn to consider how the early German romantic philosophy of nature was arguably the most important victor to emerge from the overall Controversy.

In *On the Divine Things,* Jacobi aims to express his position, which he knows is increasingly a minority one, that deploying rationalized ontologizing (as one's sole source of beliefs about what and how the real world actually is) constitutes nihilism, writing the essay as the capstone to, in Livieri's phrase, Jacobi's "lifelong endeavor to demonstrate the necessity of apprehending the object of thinking as outside the subject[88]". Jacobi writes in 1809: "The whole world is *naturalist.* I am not[89]". For Jacobi, in Livieri's gloss, naturalism has an "idealistic soul" and can never represent enlightenment or progress, but on the contrary, only "backward stepping that yields no progress, as it stumbles toward the *reductio ad absurdum* of its principles.[90]" For Jacobi, again returning to Livieri's gloss, "the assumption that the whole of reality is but the activity of an immanent principle, as if thinking on the one side and nature on the other were just echoes of a formal self-generating whole," is a "key misconception" of idealism and naturalism.[91] This idea of nature as the unfolding of a rational and intelligible principle (and ultimately, in recognition of its fundamental overall unity, only one principle—a distorted reinvention of Spinozan "necessity" itself) is an important feature of the early German romantic philosophy of nature, as we will see in the next chapter.

In *On Lichtenberg's Prophecy,* an 1802 essay of Jacobi's which he reprints as the Introduction to *On the Divine Things,* Jacobi presents a darker version of his "weaving of weaving" image from the *Letter to Fichte.* He evokes even more starkly the absence of any *meaning* in a nature seen only

[88] Livieri, A Study, 7.

[89] Livieri, A Study, 7. Emphasis added by Livieri; Livieri here quoting Jacobi and citing translation by Roth of *Friedrich Heinrich Jacobi's auserlesener Briefwechsel,* 2:414.

[90] Livieri, A Study, 7.

[91] Livieri, A Study, 7.

in the mold created by our own lower faculty of theoretical reason, our own system-building based on such a flimsy and internalist type of "necessity": such a "nature", if we are determined to imagine it as real, is like a "horrible mother-night breeding from eternity to eternity only appearance and shadowy life. And at the dawn of the day, only murder and lie[92]". This should be read differently from the more common critique of Cartesianism or causal rationality which arrive at similar normative descriptions of its outputs particularly after World War II, because these more common views accept the complete descriptive and even explanatory success of such a "world", and only fault it on moral grounds. Jacobi is faulting its content as being a horror to the self's fuller reason because it is *only a pseudo-world*—it is an understanding of the world which renders nature a "non-thing with no heart and will, an obscure shapeless being creating shapes without essence … one that in the absence of essence eternally builds images without any model, purely out of other images; a horrible mother-night"[93]—yet we accept it in place of the real; the act of *replacing* the real with this non-entity is the first source of his horror and recoiling. Wholly aside from the problem of what the causal network claims to uncover or what it claims to be "like," the very act of forcing the self to accept this network "as" the real involves a capitulation of the fuller sense of reason, the "whole inner self," in Jacobi's phrase, which is not redeemable and is even tantamount to a type of suicide. Every aspect of encountering the external real world and all the truths this encounter enables is given up for the false, but "certain," kind of truth afforded within the internalist paradigm's pseudo-world, a pseudo-world which, when stretched to ostensibly encompass all of infinite being in its monist assertion of what the real world is, becomes this dreadful, humanity-cursing "mother-night" generating only "murder and lie" as it condemns people to live only in "appearance and shadowy life"—becomes a nihil of reason which actively tyrannizes the metarational, full self, and something from which we might eventually, Jacobi presciently implies, no longer remember how to escape.

Jacobi argues for the restoration of the metarational part of our reason which can still see and understand this problem, and which is *able* to recoil and in so doing re-establish its authority over the concept-driven and

[92] Jacobi, 2024, On Lichtenberg's Prophecy, in *On the Divine,* 109 (Sec. 12).
[93] Jacobi, Lichtenberg, 109.

logic-driven sub-faculty.[94] He also invokes once more his long-running critique of the coherence of naturalism's claims to disprove God's existence: since the criterion is not actually satisfied that nature's real being is well-captured by the outputs of theoretical reason, which paint the world as only a system of knowable necessity relations or a mere "weaving of weaving", then God and nature are both released from theory's [aspiring] ontological accounts, and both, for Jacobi, may exist (and may not have their existence "disproven"). Secular naturalism has no rightful claim to the authority or the capacity to sort the existent from the nonexistent, and its efforts to call God a fiction overlook the "fabricat[ed]" character of its ontology of the real natural world; "if the human being can only have a fabricated God, then he can also have only a fabricated nature[95]". If secular (or pantheistic) naturalists wish to descry any God that will not fit the causal necessity webs of theory as therefore unreal, they need to turn the same scrutiny on the external natural world's independence from these constructed webs as well, and accept that they are thinking as untethered idealists, who seek to live inside a *nihil* of their own making, not realists in the Jacobian sense of the term.

The emptiness of what we construct ourselves, he writes in *On the Divine Things,* derives from allowing theoretical reason to arbitrarily block truly realist, metarational judgments from taking shape, resulting in a kind of self-perpetuating imprisonment of the subject away from the real:

> We know what is hardly worth knowing; we know completely and with sufficient insight only those truths and beings which, like those of mathematics, have more being and truth in their image than in the thing itself. More precisely, they are actually true only in image, and have nothing but relations and forms of such relations as their content.[96]

Science is but the systemization of everything we find we can do within this basic state of ignorance, which becomes all the more radical as we refuse to accept that it is actually ignorance. Relying on these image-truths to put us in contact with what we take to be real nature leaves us not only "powerless *with these laws* against our radical ignorance" but, worse, "rather … *distracted from it by them*[97]". Only the fuller reasoning faculty,

[94] Jacobi, Lichtenberg, 114f.
[95] Jacobi, Lichtenberg, 118.
[96] Jacobi, 2024, On the Divine Things and Their Revelation, in Jacobi, *On the Divine*, 144.
[97] Jacobi, On the Divine Things, 144, emphasis added.

the metarational person, can see and recognize this inverted state of affairs, wherein full "knowledge" can only be gained of the self's fabrications, yet in being taken for knowledge of external reality, thereby continually replaces it. The "distraction" from our ignorance which theory provides, comes because these exercises "[while] away the time"; metarational, fuller reason's "cognitions," by contrast, "do not have as their object what whiles away the time, but rather what brings it to a standstill and resolves it ... the purpose of nature and ultimate human goals, the sense of God and the essence (*Wesen*) of truth[98]". These cognitions spring up, for Jacobi, uniquely in every individual; they cannot be grafted onto one's self from without or programmed out for one, even through the rituals of religion, let alone through following the dictates and certainties of universalized, standardized theoretical reasoning and logic.

Jacobi expresses a Platonic faith in the reality of the good, the true, and the beautiful *outside* us, and our higher (or broader, metarational) reason's ability to intuit and move toward them as if by instinct, and never via proof.[99] True human reason *presupposes* this as really the case, but also does not seek to explain it through false certainties; this is in a sense part of the Thou that first forms our "I" and begins all thought and reason; without presuming truth, goodness, and beauty outside us but also intuitable by us,[100] no aspect of our reason, even our theoretical faculty, would be able to function at all. This, again, is part of the irrationality, for Jacobi, of actually thinking we might "world-make" through elements graspable to our reason and linked through necessity alone; in such an exercise, part of what would have to be initially seized "by the root so to speak, in order to throw it from[101]" the "I" as it really exists—to then permit the necessity-relations faculty to step in to the void and begin to world-build – would be this initial state of affairs which even logical reason requires as its constant frame and ground in order to function coherently. For Jacobi, there is no meaning in ideas without something truly external being the focus of our consideration, and/or acting as thought's "draw" or attractor (Jacobi seems to draw this from Plato and also from his own preoccupation with cognizing the primal reality, for the self, of the external Thou, suggestive of a dialogue or constant exchange with God and world in which the "I"

[98] Jacobi, On the Divine Things, 145.
[99] E.g., Jacobi, On the Divine Things, 149f.
[100] Jacobi, On the Divine Things, 149f.
[101] Jacobi, Edward Allwill's, 495.

is properly understood to be unavoidably situated). This makes the fabricated "worlds" of theory, which claim to offer an abstract model of reality, independent of this real exchange, and claim to place the human subject inside solely a system of factual necessities and identities known to theoretical reason alone, also disingenuous in their claimed purity (cf. Chap. 2 below), on top of all their other errors and deceptions.

Jacobi credits to Kant the "discovery that we only comprehend and perfectly conceive what we can construct to the system of identity[102]", and since Jacobi considers the false worlds of theoretical speculation to be objects made by our reason which our reason can perfectly conceive, draws support in some ways from Kant for his diagnosis that these worlds are always identitarian, and never contain more than what is put into them to originate them [logically]. He also agrees with Kant's position that there are rightfully "only two sciences in the strict and real sense of the word: mathematics and general logic," and that "all other disciplines" could "earn the title of science only insofar as they turn their objects—through a kind of transubstantiation process—into mathematical and logical beings.[103]" However Jacobi does not buy Kant's claim that Kant merely stops there, with having only "abolished knowledge in the domain of the suprasensible" and freed the world from "metaphysical dogmatism" by exposing such an illegitimate transubstantiation.[104] Jacobi reiterates his frustration with Kant for not acknowledging the chain of follow-on consequences of such an ostensibly merely delimiting exercise: Jacobi writes that "Kantian criticism, rigorously carried through to its consequences, had to have the Doctrine of Science for result, and this science, in turn rigorously carried through, had to result in the doctrine of the Unitotality, an inverted or transfigured Spinozism, an ideal-materialism."[105] Kant's approach had set up such a robust contradiction between scientific knowing and suprasensible questions that the latter ended up utterly devalued and removed from the human field, even though Kant sought to preserve them apart. The "supremacy of [practical or metarational] reason" was acknowledged too late in Kant's system, and still ended up "neutralized" and "dispossessed" by the theoretical understanding, to which Kant had granted "veto-power" over all; practical (or metarational, fuller)

[102] Jacobi, On the Divine Things, 165.
[103] Jacobi, On the Divine Things, 163.
[104] Jacobi, On the Divine Things, 164.
[105] Jacobi, On the Divine Things, 165.

reason—again, despite the fact that Kant tried to also retain it as a power of thought or intuition beyond our scientific faculty—"could not make good on behalf of faith, outside the domain of science and cognition, what the theoretical (the understanding) had destroyed on behalf of science and cognition. The doctrines of God, immortality, and freedom had to be straight out abandoned. There remained only a doctrine of nature, a philosophy of nature."[106] Even further than this, Jacobi writes, soon "not even that [philosophy of nature] remained," because eventually theoretical reason "discovered that what until then was generally called nature and its necessary laws was nothing but the human mind together with its entirely subjective representations, concepts, and associations of ideas," and "nature... became nothing".[107]

Jacobi criticizes the "doctrine of the Unitotality" (which, as Livieri points out, is Jacobi's final gloss on Schelling's system) because for Jacobi the Unitotality is a non-entity, the type of

> idol that the understanding creates for itself after having denied the superior authority of reason, lowering it under its supervision and rule to a mere faculty of fiction-making; this idol, which the understanding is now eager to raise above itself in the place of reason, invoking and deifying it, is called totality; in truth, it is only an ever-expanding universal that comes to be in the understanding through concept and word, nowhere to be found outside it—neither as one, nor as many, nor as all.[108]

The illegitimate transubstantiation of higher-reason kinds of concepts into things which fit the mold of science which Kant had outlawed (and

[106] Jacobi, On the Divine Things, 171–172.

[107] Jacobi, On the Divine Things, 172.

[108] Jacobi, On the Divine Things, 190. Cf. also Livieri, A Study, 13. Livieri also notes here that Jacobi's criticism was sharpened, this time around, by Schelling's increasing turn to transcendental idealism-infused styles of philosophizing, and that he was "far more severe" with these kinds of systems than with "scientific naturalism tout court." While I appreciate Livieri's comment, and agree insofar as Jacobi did welcome the discoveries of science as long as they were not used as a full replacement for the real, I would still counter this somewhat by noting that what most seems to animate Jacobi's critique in all three episodes of the Controversy is not the details of any one of the Spinozist philosophical-speculative "isms", but that they are pretending to *be -isms of any type*, to provide systems of true belief regarding being, while also being fully derived from the theoretical faculty's manner of thinking, dividing, and analyzing. This falsely elevates the theoretical mode of thinking over and before the metarational, inverting the correct hierarchy of the different types of reason in the subject, just as it inverts the correct order of things (moving from I to world and remaking it in the I's image, rather than moving from Thou to I) with respect to the really existing world.

which Jacobi agreed was illegitimate), was now simply taking place in reverse, and was a far worse usurping, a far more grievous sin of reason, for Jacobi. Kant, again, is held to be partly responsible, because he claimed too much primacy for that lower faculty, and asserted in effect that where metaphysics appears to inveigle itself into the realm of reason and theory, philosophy's answer must be to banish metaphysics to a separate path and keep to its certain forms of knowledge, rather than to, as it were, emphasize banishing scientific reason and theory from these metaphysical questions of great importance, and help the subject see how it must contemplate them. Kant can be seen as another, even the prime, target in some ways of Jacobi's *Letter to Fichte* cited earlier, where Jacobi opposes Fichte's type of "love for the knowledge of knowledge as such, the insight into seeing as such, the doing of doing as such." While Jacobi does not want to see divine truths forced into the mode of proofs, again, he is still diametrically opposed to Kant's view – a view shared, Jacobi assesses, by all naturalists, whether Fichte-ian or Spinozist-Schellingian or Kantian – that the theoretical, mathematical-logical faculty is somehow our most reliable part of reason for forming a complete understanding of where we are and our true condition, and that many seeming metarational truths are mere illusion insofar as they claim to be irreducible to generalizations of some truths known to this logical faculty[109]. Just as Kant believes he unmasks overreach by metaphysics, Jacobi portrays theoretical reason as also capable of extreme categorical trespass, contra Kant's implicit belief (or assumption) that lower reason will always remain self-limiting and disciplined or keep to its self-defined sphere of validity. For Jacobi, again, metarational truth-knowing and -seeking is part of our experience of the really existing world, part of really being alive, and is never reducible to operations of the mathematically rational sub-faculty, a sub-faculty which instead the metarational, fuller sense of reason must actively seek to manage, contain, and delimit, in order to prevent its nihilistic overreach which threatens the basic conditions we require (awareness of the real as external and preconceptual, and also as the source of the good, the true,

[109] Cf. on this point di Giovanni, Introduction, in Jacobi, *Main Writings*, 9. Kant, again, does try to hold ground for practical reason, including for applied morality and ethics; but this is famously unresolved in his work. Cf., e.g., Immanuel Kant, 1881, *Critique of Pure Reason, trans.* F. Max Müller, New York: Macmillan, Part II, Introduction, I.2, where Kant writes that *applied* ethics "can never constitute a true and demonstrated science, because, like applied logic, it depends on empirical and psychological principles."

and the beautiful) to be able to know anything at all, and even to make full sense of the truths of logic themselves.[110]

Despite Jacobi's repeated efforts to make his case, the Jacobian position can clearly be seen as the "losing" side of the Controversy, historically speaking; the West definitively chose the Spinozistic inheritance, and the license it found there (or put there) to ontologize – to definitively catalogue, and speculate directly upon, the really existent using logical necessity-relations alone. One group among the "winners" of the Controversy, the early German romantics, would develop a philosophy of nature which was particularly alluring to the West's new mode of ontologizing, its new naturalism and secularism, because it went through the surface motions of accommodating some of Jacobi's concerns, and may have seemed to alleviate, for those who did not look too closely at Jacobi's actual positions, some of the tensions he insisted were there.

Jacobi set up a clear either/or: either one annihilates the actually external in the name of entirely submitting its very existence to the terms of theorizable necessity, or one acknowledges our radical ignorance and the utter precedence of the really external world-as-Thou to the "I", and maintains at the head of the self the metarational or fuller reasoning capacity which alone can register and express (if imperfectly) these primary realities, and regulate and contain theoretical reason's inherently empty, and actively nihilistic, outputs accordingly. The environmental meanings, and consequences, of the historical fact that the West enthusiastically, and largely in a German romantic spirit (soon deeply intertwined with the burgeoning sciences), chose the first branch of this either/or, are perhaps already beginning to suggest themselves. I will begin to draw out these environmental meanings and consequences in more detail in the next chapter, after we familiarize ourselves with the early German romantic philosophy of nature itself. In the third chapter, I will trace a second recurrence of something like the Spinoza Controversy in between the world wars in Europe, including its implicit environmental dimensions. In the fourth and (final) fifth chapter, I will turn to discuss how the early twenty-first century West arguably finds itself immersed in the same Controversy for a third and apparently final time. In our twenty-first century iteration, the environmental meaning of the Controversy is no longer tacit, but arguably utterly central to its meaning and ultimate stakes. Given humans' rapid "vanishing" of the non-human world – which must indicate

110 Jacobi, On the Divine Things, 169–170.

some radical and basic failure of our very capacity for what we might call, in a neo-Jacobian sense, metarational environmental cognition – Jacobi's argument that naturalism was unavoidably nihilism with respect to the really existing world has never seemed more relevant. And the still omnipresent and dominant German romantic counter-argument, asserting an idea of a single knowable nature as a unity, and as a unity also incorporating human theory, art, and mind, such that the internal/external distinction could be meaningfully overcome, has never seemed more in need of radical critique.

CHAPTER 2

The Controversy's Romantic Legacy: Nature, Necessity, and … Nothing

The Spinoza Controversy in a sense both symbolized and in its small way helped further catalyze the significant shift taking place around 1800 in the European subject (again, "subject" here meant in the Jacobian sense) from an Enlightenment to a post-Enlightenment form(s). And while the post-Enlightenment, and post-Controversy, Western subject has taken many forms (and continues to evolve, as we will return to below), common to almost all of them has been both the view that the Jacobian position of circa 1800 is negligible or not as serious as the naturalist position, and, relatedly, the adherence to a certain core idea of nature (and the absence of real limits on our scientific knowledge of it) that has its origins in the circa 1800 early German romantic philosophy of nature developed by thinkers like F. Schlegel, Schelling, Novalis, and Hölderlin. Aside from Schelling, these key German romantic writers were also poets, novelists, critics, and essayists; but they are also now largely recognized as philosophical thinkers, ones whose philosophy of nature was not confined to Schelling's *Naturphilosophie*—his more formal effort to directly speculate on nature conceived of in Spinozistic terms forming but one facet of these romantics' efforts to approach nature anew.[1] During the Controversy,

[1] Again, while they differed from one another in certain aspects of their thinking and their methods, their basic epistemological approach – combining aspects of Kant's attention to the subject's role in shaping truth with other elements which moved away from Kant – and basic

K. C. Snow, *Our Inevitable Third Spinoza Controversy*,
https://doi.org/10.1007/978-3-032-03045-0_2

these thinkers were diluted among a number of other basically pro-Spinozist positions, positions with which these romantics were engaging and upon which they were seeking to build. In the roughly two centuries since the Controversy, the early German romantic philosophy of nature has arguably proven itself the most important and persistent vehicle of the Spinozistic (or neo-Spinozan–but not actually Spinozan, as we will return to below) approach to reality which in each Controversy iteration has seemingly advanced its hold over Western mainstream views of nature and vanquished the Jacobian or neo-Jacobian side.

Before the early German romantic view of nature would witness such recurrence and prominence (and further variation), it was, again, to some extent originally forged in the crucible of the Controversy, and it bears the mark of this influence. Its essential unification project among scientific and mathematical investigation, intuition, life experience, and art (or the awareness and interpretation of beauty) resembles an elegant effort to make theory capable of arabesque. Romantic "nature" was meant to evoke the one unity of all things, a unity which was yet constantly exhibiting change, growth, creativity, tension, and new forms of self-invention.

conception of a single unified infinite nature as ultimately the inaccessible absolute, but also as a Nature resembling (in their interpretation) Spinoza's *Natura naturans,* with visible, scientifically and mathematically accessible nature akin to his *Natura naturata,* was shared by all four with enough consistency that we can sketch theirs as one philosophy of nature.

In using all four thinkers as contributors to the early German romantic philosophy of nature, I am combining readings from those (e.g. Frank) who would treat Hölderlin, Schlegel, and Novalis as early German romantic philosophers but read Schelling as a German idealist, with those who would add Schelling to the list (e.g. Nassar, Rigby, and Thielke) but sometimes remove Hölderlin. Beiser also views all four thinkers as early German romantic philosophers, pointing out that Hölderlin's important role in "formulating the romantic conception of nature" included influencing Schelling (his former seminary housemate) as well as Schlegel, while Schelling, Novalis, and Schlegel all mutually interacted with and influenced one another.

Aside from Hölderlin, who left Jena after a year, the other three were based in or retained strong connections with Jena, which became an important center of early German romantic thought, while also playing host to a number of competing ideas and schools, including the idealism being developed by Fichte on the basis of Kant. Cf. Manfred Frank, 2004, *The Philosophical Foundations of Early German Romanticism,* trans. E. Millán-Zaibert, Albany: State University of New York Press; Dalia Nassar, 2014, *The Romantic Absolute: Being and knowing in early German Romantic philosophy,* 1795–1804, Chicago: University of Chicago Press; Kate Rigby, 2004, *Topographies of the Sacred: The poetics of place in European romanticism,* Charlottesville: University of Virginia Press, 34; P. G. Thielke, 2010, Who's Who from Kant to Hegel II: Art and the Absolute, *Philosophy Compass,* 5: 408n26; Beiser, *German Idealism,* 397, 393, 467; Safranski, *Romanticism,* 105.

Drawing from their own, and various of their contemporaries', understanding of Spinoza, they sought to portray contemplation and experience of this single unified nature as a way for theoretical understanding to move out of the shadow of the dated worldview of the Enlightenment and even partly overcome Kant (cf. also Beiser on this point, per note 12 below), while also bearing Kant's critiques in mind. That this philosophy of nature was undertaken partly as a rejoinder to Jacobi's skepticism and seeking to overcome it in a sense only spurred its arabesque leaps ever higher.

We will consider in what follows next the basic outlines of three main elements of this philosophy of nature most relevant in the context of this work: first, its manner of approaching the connection between the finite thinking subject and the infinite universe; second, its new way of combining epistemological and ontological claims; and third, its overall aesthetic self-interpretation project these thinkers called their efforts to lay the foundation for a "new mythology of reason." All three elements of the philosophy, and the basic outlines of the "nature" it aims to represent and encompass, will reappear in both the post-WWI and our own early twenty-first century Spinoza Controversy landscapes. In the later Controversy iterations, neo-romantic "nature" becomes the preserve of the natural sciences, foundationally physics and cosmology. Here, in the context of the original Controversy, the romantics present their claims about Spinozistic nature through poetry, fiction, art and historical criticism, and through philosophical fragments and essays, except for in the case of the prolific Schelling, the most formal philosopher of the group, who churned out voluminous quantities of traditional long-form philosophical texts and lectures.

At the time of the original Controversy, these romantics themselves seemed to see their turn to nature, and their construction of a particular way of experiencing the act of knowing it, as a potential third way out[2] which incorporated elements and critiques from both sides of the debate. Jacobi was relentless, for his part, in insisting on the either/or noted at the end of the last chapter. Necessitarian, theoretical reason must either pursue, or demur from, claims to engage in speech of being itself or the really existing world in full or as such. The new early German romantic version

[2] On this "third way," would-be peacemaker, component of the romantic turn to nature in the context of the Spinoza Controversy and especially on Herder's offering them a particularly amenable version of neo-Spinozism with which to work for these purposes, cf. also Lamm, Romanticism, 167ff.

of Spinozistic nature which emerged from the Controversy is in a sense even more nihilistic, from a Jacobian point of view, than unadorned Spinozan thought, because romantic philosophy's claimed union with art and the aesthetic sense provides its concept of nature with a more convincing externalist mask, one that may further conceal its primary allegiance to [always internalist] theoretical reason from those who do not look carefully.[3]

Early German Romantic Philosophy of Nature: A Key Response to the Controversy

What would come to be known as the early German romantic movement (*Frühromantik*) is usually considered to have begun around 1794 and to have ended around 1808.[4] The first bookend marks the year Schelling and Hölderlin spent together in Jena, where Schlegel and Novalis would soon form their salon. The other bookend, 1808, is placed when – seven years on from his close friend Novalis's untimely death in 1801, and two years after Hölderlin, in worsening mental health, had stopped actively writing – Schlegel converts to Catholicism, with Schelling also by this point beginning to gravitate away from his earlier romantic approaches to, and focus on, nature.

From among the wide-ranging Anglophone secondary scholarship on these four primary early German romantic philosophers of nature, my own reading of their philosophy as cosmological and ontological (and directly Controversy-shaped) draws some support from, for example, Beiser, Nassar, Thielke, Gare, and at times Bowie (who does not classify Schelling as a romantic but at times seems to ascribe to him romantic positions). Nassar's *The Romantic Absolute* (2014), in particular, helpfully draws attention to the way in which epistemology and ontology cannot be separated in these romantics' thinking, as their efforts to think nature as the absolute required the creative intertwining of them. Beiser, in works like *German idealism: The struggle against subjectivism, 1781–1801* (2002), and *The Fate of Reason* (1993), offers a reading of these romantics as even

[3] Given space, it could be instructive to consider how later literary and artistic criticism of various of the romantic writers and artists often makes the same observation, albeit from another angle.

[4] Cf., e.g., E. Millán-Zaibert, 2005, The revival of *Frühromantik* in the Anglophone world, *Philosophy Today*, 49: 102.

more heavily engaged in metaphysics and ontology than they are in Nassar's view. Gare reads Schelling's *Naturphilosophie* as an effort to try to move philosophy as a whole from thinking from the standpoint of Kant's too-isolated subject, the "standpoint of reflection," into thinking from the "standpoint of production":

> Referring to ... the 'standpoint of production' in contrast to the Kantian 'standpoint of reflection', Schelling was concerned not only to show the social conditions for objective knowledge, but the nature of the world that enables it to be known objectively and explained at least partially through Newtonian physics, while [also somehow enabling the subject to] question ... the assumptions of Newtonian physics. ... [He wanted to explore] the development of a new physics which he claimed would reveal the relationship between magnetism, electricity, and light, provide the theoretical foundations for chemistry, justify and advance Kant's conception of life, and provide a new way of understanding human existence. This, in essence, is the whole project of Schelling's *Naturphilosophie*.[5]

This quest to occupy the standpoint of production, rather than that of isolated Cartesian or Kantian reflection, seemed to animate all four thinkers. These romantics saw thinking as always unavoidably pre-grounded in being or nature, a situation that could only be intuited and taken as a prior given, not represented conceptually[6] – one of their efforts to respond to Jacobi's claims about the unavoidable grounding of all thinking in Thou-cognition. This pre-grounding in being added to the always-elusive character of the highest goal of learning to think from Nature's own standpoint of production, "making" nature. The subject could, however, *approach* this stance intuitively and via leaning on a monist conviction of the ultimate unity of mind with nature—and ultimately also with Nature—while striving to see nature primarily *as a product of* ungraspable Nature.[7] This monistic Nature-nature-mind continuum used neo-Spinozistic conceptions, reworking (and distorting, or at least reinventing) into an empirical

[5] Arran Gare, 2018, Natural Philosophy and the Sciences: Challenging Science's Tunnel Vision, Philosophies, 3 (33): 18 of 29; Gare here cites F. W. J. Schelling, *First Outline of a System of the Philosophy of Nature* [1799], 2004, Trans. K. R. Peterson, SUNY Press: Albany.

[6] Gare, 2018, Natural Philosophy, 17 of 29. Cf. also D. Nassar, 2011, 'Idealism is nothing but genuine empiricism': Novalis, Goethe, and the ideal of romantic science, *Goethe Yearbook*, XVIII: 80f., re Novalis on this point.

[7] Gare, Natural Philosophy, e.g. 17–18 of 29. Cf. also Andrew Bowie, 1993, *Schelling and Modern European Philosophy: An introduction*, London: Routledge, 13, 35.

mode Spinoza's famous *Natura naturans-natura naturata* relation[8] and his framing of the absolute entirety of being as somehow intelligible, while enjoining the subject (in a hallmark of the romantic way of thinking nature) to utilize all forms of reason, intuition, and artistic capacity, unified into one multifaceted [meta-] faculty or mode of sense-making, to understand the picture of reality which resulted. The weight the romantics placed on occupying the imagined standpoint of production also led them to emphasize the importance of learning from, and using as some foundation for truth, the sheer experience itself of rational thought (foreshadowing the logical empiricists in the post-WWI period) as well as the sheer experience of making art. The production of art could also, for these thinkers, be seen as a philosophical act of knowing the world through

[8] The *Natura naturans* (Nature naturing) – *natura naturata* (nature as "natured" or "made" or "produced" by Nature) distinction/unity is one of Spinoza's concepts that greatly interested the romantics, especially Schelling, as we will see in what follows. In many studies of how Spinozan ideas were used in early German romanticism, the pairing is used together, as "Nature-nature," to represent how the romantics understood and explored the concept. The romantics' de facto single-case configuration of a Nature-nature (whereas for Spinoza Nature had an infinite number of attributes (and modes), of which thought and extension themselves, in their entirety, were but two, as we will return to below) toward which we were able to directly empirically orient ourselves as knowers is of course quite different from the way in which Spinoza used the terms. In his *Ethics,* Spinoza writes:

> … what we should understand by nature viewed as active (natura naturans), and nature viewed as passive (natura naturata) [is:] by nature viewed as active we should understand that which is in itself, and is conceived through itself, or those attributes of substance, which express eternal and infinite essence, in other words … God, in so far as he is considered as a free cause.
>
> By nature viewed as passive I understand all that which follows from the necessity of the nature of God, or of any of the attributes of God, that is, all the modes of the attributes of God, in so far as they are considered as things which are in God, and which without God cannot exist or be conceived. (Spinoza, *Ethics*, Note to Prop. XXIX).

All citations of Spinoza's *The Ethics* in this work are taken from B. Spinoza, 2009 [1677], *The Project Gutenberg E-Book of The Ethics*, https://www.gutenberg.org/files/3800/3800-h/3800-h.htm. Scruton emphasizes that "[t]here is no contradiction between these two ideas [of Nature/natura naturans and nature/natura naturata]; each approaches the one identical reality, from two contrasting points of view." Cf. Roger Scruton, 2002, *Spinoza: A very short introduction,* Oxford: Oxford University Press, 52–53. Arguably Spinoza's dual term is primarily designed to allow him to defend his monism, and explain how there can be an appearance of a variety of types of being without there being in actuality, for Spinoza, more than a single real substance.

remaking it, and insights drawn from thinking about art (as well as from their own experience creating literary art, for all except Schelling) occupied a central role in their philosophy, informing what constituted viable evidence of having reached ultimacy in human knowledge, including, in particular, human knowledge of or insight into nature. In their view, if art could bring us an incontrovertible glimpse of the absolute and our own grounding in it (even as it reminded us ironically of our own limits), then science and mathematics must also be somehow art-like. Both science and art must target truths about the one unified [romantic version of] Spinozistic Nature-nature, thus mathematics and science must learn to reveal something beautiful in nature that resembled artistic production, just as they must be accurate and with this accuracy help furnish the new symbolic language conveying the one unity (cf. also note 99 below). Science must be essentially somehow one with art; accuracy must be one with ultimacy; else the subject must be in error in both realms. The equivalency could be ironically experienced or cause despair, but it could not be denied. For Jacobi, such an equivalency between accuracy and ultimacy was neither elevating nor tragic but instead nonsensical, as if the subject, drowning in a barrel, forgot they had elected to put their head into it and lift their feet, and could simply stand upright whenever they chose to do so.[9]

The romantics, for their part, did all they could to answer Jacobian objections to the reality of their single "nature" by romanticizing—by elevating, infinitizing, and transforming it through the lens of seeing it as an artistic product of the unseen God or Nature—this nature's ostensible

[9] While this metaphor is my own, Jacobi played often with spatial and embodied metaphors comparing substitution of one's own abstract system for the real world to trying to walk on one's head. In this false position, the human being tries to navigate the realm of real experience ("walk") using the abstractions of reason (rather than our feet, which are built and molded for experience). Jacobi famously called during the Controversy for naturalists to make a rapid *salto mortale*, or redeeming or existential leap or jump, to right themselves and return to sanity and reason. Jacobi's *salto* is sometimes misinterpreted as "leap of faith", but actually refers to a need for humans to stand upright once again, to restore our metarational reason hierarchically above and over our theoretical, precise, certainty-based world-building faculty, and to cease believing that theory's world can encompass the real one as such, or cover the realm of real lived experience. He also seemingly resorts to the image to convey the completeness of his opposition to the new uses to which Spinoza or Spinozism was being put and the holistic quality of the errors of naturalism when seen from the metarational perspective. We will return to this point in Chap. 5. Cf., e.g., Jacobi, Concerning the Doctrine, 194–195, and di Giovanni's footnote on 195, where he notes that "the expression 'leap of faith' is nowhere to be found in Jacobi."

form and content qua knowable and beautiful single unity inclusive of the human being. These romantics, under the influence of especially Herder, favored more "*living*, or organic" scientific paradigms for nature than they imagined possible in the Newtonian conception; nature for these thinkers was ultimately constituted of a single force, as Forster notes, and they were keen on carrying forward and refining the spirit of Herder's "metaphysical-religious monism" that had "swept through German philosophy" in the late 1700s and early 1800s, counting Schelling, Schlegel, and Novalis among its adopters,[10] and fueling and intertwining with the Controversy, as noted in the first chapter. Bell notes that Herder's mis-reading of Spinoza's system as somehow directly empirical or able to be incorporated into the practice of physical science was bequeathed to these romantic thinkers.[11]

In Beiser's view, the romantics embraced Spinozism because the "old theism had collapsed to the forces of Biblical criticism; and deism had faltered in the face of Humean and Kantian criticism. Only Spinoza's pantheism was not vulnerable to rational criticism or scientific advance[12]". As we will continue to see, the romantics' embrace and elevation of the concept of necessity was key for why and how [their version of] Spinozism offered a view of nature and our ultimate knowledge of it which was ostensibly so broadly and durably defensible, including in eras well after their own.

In Beiser's account, as well as for Crowe referencing Novalis, and Frank in reference to the evolution of all four thinkers except for Schelling,[13] the Spinoza Controversy and Jacobi exerted a formative influence on the early German romantics in that they strove to avoid sharing Jacobi's main conclusion – essentially, the need to forego using theoretical naturalism to make claims in a Spinozist manner about really existing nature – while agreeing with him on other points, e.g., the utter prior-ness of being to the processes of conceptual reason (as noted above), or, as Nassar points out, Jacobi's criticisms of Kant's inaccessible thing in itself as an incoherent concept.[14] Hölderlin perhaps remained the closest of the four to

[10] Michael N. Forster, 2012, Herder and Spinoza, in *Spinoza and German Idealism*, 79.

[11] Cf. Bell, *Spinoza in Germany*, 111.

[12] Frederick Beiser, 2006, The Paradox of Romantic Metaphysics, in *Philosophical Romanticism*, Ed. N. Kompridis, London: Routledge, 226.

[13] Cf. e.g. Beiser, *German Idealism;* Benjamin Crowe, 2008, On 'the religion of the visible universe': Novalis and the pantheism controversy, *British Journal for the History of Philosophy*, 16: 125–146; Frank, *Philosophical Foundations.*

[14] Cf. Nassar, *Absolute*, 5: 92, 96, 120.

Jacobi,[15] but even Hölderlin does not entirely forego monistic speculation, and often evokes absolute mind-nature unity in his writings, as well as the capacity of the human subject to serve as nature's direct continuation and culmination, without encountering any gap in kind or anything Other to self. In his novel *Hyperion,* the title character speaks to nature as a man speaking to a woman in a way which gives "her" no true Thou or primal Other status: "'They will come, your men, nature! A rejuvenated people will rejuvenate you too, and you will become as its bride, and the ancient covenant of spirits will renew itself with you. 'There will be but one beauty; and humanity and nature will unite into one all-embracing godhead.'"[16] The position was held by all four romantics, in varying ways, that one must and could begin to think nature, as Schlegel writes (and as Nassar highlights was particularly important in Schlegel's thought[17]), "in medias res[18]", avoiding speculating as to raw origin-points or beginnings, beyond metaphorical ones used to explore constitutive essences and processes in nature (and link these with the thinking "I" as well). Again, this situating of thought within pre-existent reality can be read as a significant concession to (or shared common ground with) Jacobi. Yet I would argue that by Jacobi's most important criteria, such concessions were no more than aesthetic maneuvering. The romantics' deep-running monism still ended up neutralizing most such gestures toward Jacobian reservations regarding theorizing real being. The romantics acknowledged that philosophy, both in its Enlightenment form and in the forms being developed by themselves and their contemporaries, was deeply flawed as a way of understanding nature and properly assessing the reach of our own knowledge. But unlike Jacobi, and in a sense also reacting against Kant, their response was not to focus on better understanding these flaws or limits, but instead to seek ways to elevate and expand the theoretical-speculative philosophy at our disposal for thinking nature, including via transforming natural philosophy (the older category of nature study, resembling science) into actual intentional nature-philosophy or philosophy of [Nature-]

[15] Per Beiser, Hölderlin was influenced by both Spinoza and Jacobi, cf. Beiser, *German Idealism,* 379.

[16] J. C. F. Hölderlin, 2009, *Hyperion, or the Hermit in Greece,* Trans. H. Gaskill, Cambridge: Open Book Publishers, 77.

[17] Nassar, *Absolute,* 93f.

[18] Schlegel wrote: "Viewed subjectively, philosophy, like epic poetry, always begins *in medias res.*" Friedrich Schlegel, 1991, *Philosophical Fragments (1798–1800),* trans. Peter Firchow, Minneapolis: University of Minnesota Press, 28 (no. 84).

nature, thereby raising its relative importance. As we will continue to see, the romantics sought to develop a philosophy of nature capable of framing a new quasi-secular, quasi-pan[en]theistic, naturalism able to serve as a font of real, complex, and multi-dimensional beliefs about the physical world and humans' place within it. They auditioned a number of tools that might be placed at the new worldview's disposal, never deciding on a fixed outcome, but tending to cluster around recurring ideas with which they experimented, all oriented toward equipping humans to pursue the new elusive goal of all theoretical and artistic knowledge, which was direct and unalloyed knowledge of the absolute – something akin to the unconditioned, to God or Nature considered as the indefinable source of everything conditioned[19] – even as they simultaneously held this to be impossible for a finite human mind to sustain beyond indirect insight, insight which must be obtained and tested via nature, and via human knowledge, life, and art, since the world itself was now seen as actually "participating" in (and as being one with) the absolute, in Nassar's phrase,[20] and since revelation was no longer a viable path to ultimate truth.

Again, one of the recurring ideas with which the early German romantics experimented was that mathematics and science should be interpreted afresh, including via uncovering what must be their underlying unity with art. Despite lingering received stereotypes of romanticism as somehow anti-science, especially for Schelling, Schlegel, and Novalis, mathematics and science formed an important part of the way in which human beings could learn to perceive the physical world as part of one unified "nature" which was also the product of the activity of a quasi-divine Nature. Richards notes that "Many of the Romantics might … be classified as hyperbolically rational in that they believed the scientific mind could penetrate into all the dark corners of the universe[21]". The early German romantic thinkers speculated on the extent to which the whole of nature, conceived as the product of Nature, might be reflected or revealed in science and mathematics, now presented increasingly (at least aspirationally) as a neo-revelatory practice, particularly once science updated itself from using Newtonian paradigms for nature and replaced them with the new

[19] Schelling writes: "Nature has unconditioned reality, a proposition which is precisely the principle of a philosophy of nature." Schelling, *First Outline*, 16–17 ("First Division", nos. 81–82). Cf. also Nassar, *Absolute*, e.g. 5f.

[20] Cf. Nassar, *Absolute*, 5.

[21] Robert J. Richards, 2002, *The Romantic Conception of Life: Science and philosophy in the age of Goethe*, Chicago: University of Chicago Press, xvii.

organic and dynamic conception of nature as consisting of force and energy. One day, even if only in the distant future, the unity of all things – even though such knowledge in itself must ultimately entail knowledge of the absolute – should itself be describable and graspable via the logical part of reason, which as part of fulfilling this destined role would also show itself to be in perfect harmony with, and of the same ilk as, the human instinct for, perception of, and ability to make (in art), beauty.

Schelling, in particular, was very engaged with theorizing the process of science and even experimenting with ways one might change some actual theories within it. His *Naturphilosophie* is usually considered a kind of speculative physics (intended to in effect idealistically frame or ground the experimental side of physics). Schelling also experimented with new ideas in chemistry and biology. For Schelling there was a single continuum connecting all the ceaseless phenomena of nature, passing from connections describable using the mechanical laws up to those of the universe's "higher" dynamics, "whose spirit," Schelling writes, "is expressed in the principle that views the dynamic as the single positive and original aspect of Nature, the mechanical only as the negative and derived aspect of the dynamical[22]". Experimental science was important because it seemed to potentially offer empirical proof of the existence of the unity among all things and between mind and nature. As Peterson writes, Schelling was interested in how "experimentation would not be possible without the employment of an idea of Nature," an idea of a single unified whole of reality, and also in how this "idea [of Nature] itself" likewise must be in some way true if it could really function in such a role as usefully grounding science and experiment.[23]

Mathematics was an important key to nature for Novalis and Schlegel because it could display self-grounded progression, suggestive of a life-like, self-contained creative process; for these thinkers, the fact that mathematics seemed to be able to parallel or resemble human "cognition" seen as the same kind of "poiesis"[24] further suggested its presence or groundedness in nature as well. Physics, for Novalis, was "*real* mathematics", because "Nature incessantly adds, subtracts, multiplies, raises to a higher

[22] Schelling, 2004, *First Outline of a System of the Philosophy of Nature*, trans. K. Peterson, Albany: SUNY Press, 192 (no. 271, "Third Division").

[23] Keith Peterson, 2004, Introduction, in Schelling, *First Outline,* xxiv.

[24] Bianca Theisen, 1998, Romantic Myths of Myth: Myth as Autopoiesis, in *Myth and the Making of Modernity: The Problem of Grounding in Early Twentieth-Century Literature*, eds. Michael Bell and Peter Poellner, Amsterdam and Atlanta: Rodopi, 20.

power etc." and "The applied mathematical sciences show us Nature as a mathematician[25]". Schlegel wrote that "dynamics is the macrology of energy which, in astronomy, is applied to the organization of the universe. To that extent both could be called historical mathematics[26]".

Novalis and Schlegel were interested in the potential use of mathematics to unify truth and knowledge across not only all the sciences, but all domains of thought. Novalis wrote that "Mathematics are capable of endless perfection, as a convincing proof of the sympathy and ideality of nature and intelligence[27]", and he held that "All sciences should become *mathematics*" due to mathematics' expression of the "true scientific spirit", and that all fields – poetics, grammar, physics, philosophy, history – must each [implicitly, "already"] have their own form of mathematics.[28]

At the time of his early death in 1801, Novalis was working on an "encyclopedistics" method which would allow him to collate all fields' findings, then locate their unities via integrating (in a quasi-mathematical sense) up from their specific granular claims to derive overarching shared truths. The resulting encyclopedia would constitute the single, continually expanding, book of all [worthwhile] human knowledge. The mind's intuition of beauty, and the correctly attuned intellectual intuition—the emphasis on intuition again a sign of Jacobi's as well as Spinozism's influence—were both indispensable for, and to be highlighted and celebrated in, such a wide-ranging project. Nassar notes that this was to be an "organic whole" corpus of knowledge expressing "relations among ideas";[29] Rigby sees it as part of Novalis's effort to "disclose the underlying unity of the cosmos[30]". Novalis also dreamed of understanding individual human lives in the same way, writing in his *Notes for a Romantic Encyclopedia* that "Absolutization—universalization—classification of the individual moment, of the individual situation etc. is the real essence of romanticizing[31]". Again, while the German romantics at times might have used some of Jacobi's own skeptical language, and their own articulation

[25] Novalis, 2007, *Notes for a Romantic Encyclopedia*, Ed. and Trans. David Wood, Albany: State University of New York Press, 195 (nos. 9 and 11 in Mathematics Notebook).

[26] Schlegel, *Fragments*, 92 (no. 445).

[27] Novalis, 1891, Thoughts on Philosophy and Physics, in *Novalis: His life, thoughts, and works, trans.* and Ed. M. J. Hope, Chicago: A. C. McClurg, 230.

[28] Novalis, *Notes*, 195 (nos. 9 and 11), emphasis original.

[29] Nassar, *Absolute*, 5.

[30] Rigby, *Topographies*, 141.

[31] Novalis, *Notes*, 14 (no. 87).

of something like the ultimate "opacity of Being[32]", in Millán-Zaibert's phrase, to soften their absolutism (just as Jacobi at times also used his grasp of Spinoza and the new Spinozistic approaches to bolster and illustrate his case against naturalism), the romantics were nonetheless still taking steps toward arguing that eventually it must be possible for the human to enlist all of reason (including our art-making and our critical capacities) to elevate human knowledge of really existent nature toward the level at which we could understand it as the absolute, and this was unavoidably a [new] form of epistemological and ontological absolutism (we will return to their views of the absolute below). In the romantics' nature, their single unity which contained all, there were no real incommensurables, no real differences in kind, or in kinds of meaning (with all the fields of knowledge and even art and science as a whole and as such, fundamentally unified). Millán (agreeing here with Beiser) characterizes Schlegel as subscribing to a "holism that acknowledges not only a world of mechanical laws, that is, a realm of static, unchanging laws of nature, but also a world of processes of life, of genetic or organic laws[33]"; both were potentially knowable facets of the same unity, [these romantics' version of] Spinozistic Nature-nature.[34]

The Romantic Linking of the Finite with the Infinite

As noted above, the early German romantics were interested in somehow, however imperfectly, developing truth claims about infinite Nature-itself as real and as really the ultimate character of the nature that was (in their view) immediately accessible to us (and seemingly finite). Developing a meaningful way of speaking about the presence of the infinite in the finite, and vice versa,[35] was a recurring preoccupation. Schelling repeatedly describes the seemingly finite physical world as being the limit of the infinite "unconditioned reality" or absolute, and thus also infinite, as is

[32] Millán-Zaibert, Revival, 97. She notes that her reading of Pinkard is that for Pinkard, the Early German romantics "accept the opacity of Being", contra Hegel.

[33] Elizabeth Millán, 2007, *Friedrich Schlegel and the Emergence of Romantic Philosophy*, Albany: State University of New York Press, 143.

[34] Cf. Footnote 8 above.

[35] Nassar, *Absolute*, 2ff., observes that "What is the relation between the infinite and the finite?" was one of the early German romantics' three main questions or areas of interest, the other two being "What is the relation between mind and nature?" and "What is the relation between the one and the many?". Cf. also Nassar, *Absolute*, 260.

also a recurring notion in Schlegel. In situ, in romantic thought, this offered a panentheistic or pantheistic-tending quasi-religious conception of nature, including an implied type of secularized (and knowable, or at least rationally theorizable) continuous creation which was anathema for Jacobi.[36] These romantics' preoccupation with infinite-finite relations (and unity) also shaped how they sought to "infinitize" human thought itself. Schelling writes that "The *I* think, *I* am, is, since Descartes, the basic mistake [*Grundirrtum*] of all knowledge; thinking is not my thinking, and being is not my being, for everything is only of God or the totality[37]". Nassar characterizes Schelling's approach as "grounding philosophy in Being, or *Dasein*, understood both as 'that which is' (nature) and as 'existence' (implying the human subject)[38]". The other three thinkers were also prone to explore this new way of positioning the thinking "I" as grounded in a nature which was infinite (since it was ceaselessly emanating from Nature) but with which the "I" also knows it is *necessarily* somehow one. Where Jacobi found cause for radical humility in the utter prior-ness of the world to thought, these romantics found evidence that the infinite itself in some sense must think via the human. The romantic avoidance of puzzling through creation accounts or raw beginnings also ended up asserting a kind of unity and even direct cognitive contact between the finite thinking "I" and the infinite in a cosmological sense. The nature one could simply reach out to and begin theorizing and knowing (without puzzling over a creation story in the Biblical sense) was now taken as always having been in existence in some way, drawing from Spinoza's famous formulation of the "eternal and infinite Being, which we call God or Nature" (*Ethics*, IV/Preface). Yet growth and development were also held to characterize nature's changes through time, suggestive of chronological progression and thus finitude (and tractability to logical analysis). This was in keeping with their Nature-nature distinction, and foreshadowed the Einsteinian cosmology which would arrive a century later. The Nature-

[36] Cf. Jacobi, Concerning the Doctrine, 370–373 (nos. 408–419).

[37] Friedrich Schelling, 1860, Aphorismen zur Einleitung in die Naturphilosophie [1806], *Aus den Jahrbüchern der Medicin als Wissenschaft,* F. W. J. von Schelling's *Sämmtliche Werke*, Part I, Volume 7, Stuttgart and Augsburg: F. G. Cotta, 148 (no. 44). Translation in and by Bowie, cf. Andrew Bowie, 2016, Friedrich Wilhelm Joseph von Schelling, *The Stanford Encyclopedia of Philosophy,* ed. Edward N. Zalta, NPN.

[38] Dalia Nassar, 2020, An 'Ethics for the Transition': Schelling's critique of negative philosophy and its significance for environmental thought, in *Schelling's Philosophy: Freedom, nature, and systematicity,* ed. Anthony G. Bruno, Oxford: Oxford University Press, 241.

nature distinction, in the cosmological frame, apparently gave the romantics a way to combine an a priori assumption that we must always be able to know nature as progression, growth, and creativity or self-production, with an a priori assumption that Nature has always existed. Wilson discerns a romantic cosmology in Schelling in which the universe is held simultaneously to be both "physically infinite ... [and] metaphysically finite" conceived of in itself, and only becomes seen as sheer "infinite, metaphysical Nature" when viewed as the Absolute.[39] With the Nature-nature distinction (as they understood it) available to bridge the contradiction, at the cosmological scale the romantic subject thus gained an ability to know something infinite as though it were finite, yet without relinquishing the claim of the ultimate absolute infinitude of the target of its knowledge (cf. also Chap. 3). Intellectual intuition was often invoked to describe the situatedness of the finite "I" directly in and of the infinite (and the absolute) and thus able to hold this creative tension in mind (or know it via romantic dialectics). The romantics' reliance on intellectual intuition was important for their efforts to claim they could avoid some (but not all) of Kant's limits on the subject's knowledge of nature, and was another instance of their efforts to incorporate, co-opt, or otherwise respond to Jacobi's positions. For Jacobi, it was impossible to disguise the internalism of their entire construct, and the reason it led to contradictions was because it was not actually true of the real world (cf. below). But they were undeterred (at least during their romantic period; Schelling and Schlegel would later move beyond these positions). Schlegel held, in a view shared with Schelling, Hölderlin, and Novalis, that "Nature, the universe, can be felt and conceived of without mediation[40]". And at the same time, as we have been discussing, part of thinking nature directly "without mediation" *had* to entail somehow directly thinking of or as the infinite. Schlegel relished this paradoxical contact (during this phase of his thought); Nassar notes that he saw infinitude[-as-such, or the absolute] and consciousness as being not just in direct contact, but as *solely* definable via their interplay;[41] for Schlegel "Only in relation to the infinite is there meaning and purpose; whatever lacks such a relation is absolutely

[39] Andrew D. Wilson, 1993, "Romantic cosmology," in *Encyclopedia of Cosmology*, ed. Norriss S. Hetherington, New York: Garland, 601.

[40] Schlegel, *Fragments*, 98 (no. 44).

[41] Cf. e.g. Nassar, *Absolute*, 156.

meaningless and pointless[42]", and this encompassed his view of how we must weigh and understand both human existence and nature.

Again, just as the romantics posited that thinking began within being, within Nature, which was infinite, and thus must somehow be unified with it (even if it was difficult to represent as such directly), so they got around the problem of thinking the infinite itself as a concept by claiming to speak about the infinite mostly as it [ceaselessly] made itself manifest in (and produced) the seemingly finite. Schelling called "the exhibition of the infinite in the finite" "the highest problem of all systematic science," a question which "subordinate sciences" can address in the particular, but which philosophy must solve "in its greatest universality[43]". Schlegel asked, "'*Why has the infinite come out of itself and made itself finite?* – that is in other words: *Why are there individuals?*—Or: *Why does the play of nature not run itself out in an instant, so that nothing would exist?*'[44]". Taking such questions as valid ones for "naturalistic" arbitration—and in so doing tacitly accepting the romantic version of nature and what our most profound knowledge of it must look like—would prove to hold recurring allure for Western thought from this point forward, as we will continue to see in later chapters. For the romantics, sometimes this seeking of the infinite in the finite did cross into trying to grapple with the sublime or the chaotic in nature. Yet even here, where the "opacity of Being" (cf. above and note 32) seemed to present the most challenge to the thinking "I", they did not abandon the attempt to think through how even in such conditions, the "I" could at least know the underlying truth that what we found in nature must be the infinite making its appearance as the finite world, and that we ourselves must be one with it and thus somehow able to grasp it at least in principle.[45]

[42] Schlegel, *Fragments,* 94 (no. 3).

[43] Schelling, *First Outline*, 15 (no. 79).

[44] Schlegel, from around 1800, in a Jena lecture given by Schlegel on philosophy, quoted in John H. Smith, 2014, Friedrich Schlegel's romantic calculus, in *The Relevance of Romanticism: Essays on German Romantic philosophy,* ed. Dalia Nassar, Oxford: Oxford University Press, 248, citing [KA 12, 39].

[45] Later romantics, particularly some of the English romantics, would in a more Kantian way associate the sublime with where the intelligible completely broke down, as in Byron's remark upon looking through a Herschel telescope in 1811 that "'our pretensions to eternity might be ...*over-rated*'", or Shelley's *Mont Blanc,* where Oerlemans notes that Shelley calls for the human subject to recognize the "ephemeral" quality of any "legislation" over nature like Linnaeus's since what the mountain reveals is the "complexity, the infinite density, of the physical world [that] is its 'mysterious tongue/ Which teaches awful doubt' – doubt, for

The imperative to infinitize the finite (and finitize the infinite) also reflected the romantic belief that, again at least in principle, one should be able to somehow collate and then integrate the infinitesimals of a human life up to the level of the one single unity, as where Novalis speaks of absolutizing the "individual moment... the individual situation etc." (cf. above), or when Schlegel played with mathematical metaphors to contemplate this goal.[46] At times, as touched upon earlier, this interchange between the finite subject and the infinite could also lead to suffering and existential despair; Novalis writes, "We *seek* the absolute everywhere and only ever *find* things[47]"; we can feel utterly cut off from what we simultaneously know must, again, *necessarily* be the ultimate unity, one with our art and our science and our own lives (if and when they are authentically or correctly oriented toward it). Rather than interpreting this as a signpost to turn back, or to decide that an abstract concept like the absolute (or the one, the totality, the All, etc.), if linked with necessity and a theorized world which was simultaneously held to be our only reality we could experience, might lead quickly to nonsensical quicksand (as Jacobi might argue), the romantics instead interpreted this feeling of exile itself empirically, situating human beings as travelers through a world of [cognitive or perceptive] disunity and finitude who still (and once again, *necessarily*) know ourselves to be fundamentally one with a unity which is infinite. Our fate is thus to constantly seek reunion with it without ceasing, through knowledge and through art, never fully finding it but elevating our lives, potentially infinitely, by the attempt.

Schlegel's views of the infinite had this normative, moral as well as vocational quality to them; as Millán notes, Schlegel's vision of monistic reality, and opposition to Kant's thing in itself, was indicative of his "desire to free us from the illusion of the finite" world of the thing-ly altogether, moving instead toward a notion of monistically conceived infinite and

instance, about our ability to perceive a totality, a complete order". For the Byron reference cf. Paul Duro, 2010, 'Great and Noble Ideas of the Moral Kind': Wright of Derby and the Scientific Sublime, *Art History*, 33: 665; for the Oerlemans and *Mont Blanc* reference, cf. Onno Oerlemans, 2002, *Romanticism and the Materiality of Nature,* Toronto: University of Toronto Press, 146.

[46] Cf. Smith, Schlegel's calculus, 249f.

[47] Novalis, 1997, Miscellaneous Observations, in *Philosophical Writings,* ed. and trans. M. M. Stoljar, Albany: State University of New York Press, 23 (no. 1).

ceaseless change and life[48] [suggestive of what would later be called structural realism]. Freeing us from the thing-ly, in some way simply declaring the world we encounter to be really less real than, or at least literally made by or derived from, ideal, intelligible essences (even where these essences are only laws of change), is a recurring theme in the anti-Jacobian position, and will become deeply intertwined with the sciences by the time of the later iterations of the Controversy. It also gives rise in the German thinkers to a new romantic way of defining the highest and truest way of living as the undertaking of enough activity – activity of the artistic, or other abstract and intellectual, type – to *make* the ideas and outputs of logic and theory, and art, both real—even more real than what we at first see around us—and really a pathway to completely romanticizing human life, increasing its absolute significance (even when this significance is experienced negatively, through new depths or types of cognitive and artistic despair, or sense of shortcomings, which become possible in the romantic frame). For Hölderlin (as we will touch on below), the striving to become more infinite, through art and thought, is a constant creative activity which requires the human to bring nature into himself, and place himself in nature, to allow an almost plastic co-molding.

The Romantics' Quadratic "Nature": Epistemologically-ontologically delineated and ideal-real

While these four thinkers can conceivably also be interpreted as preserving slightly more Kantian wariness of direct speech of the "whole" of nature grasped as a single entity and unity than I have begun presenting here, emphasizing the presence of [at least the rudiments of] just this totalizing type of cosmological and ontological speech of nature is both arguably a stronger reading of these thinkers for their own sake (and I also draw upon, for this assessment, a number of specialists, as noted above) and also vital to lay out and understand given the particular forms of essentially German romantic ideas of nature that would come to propagate outward from this original tumultuous period of post-Enlightenment upheaval, both via direct historical influence chains and through seeming simple recurrence. With time, subtleties which were present in the original German romantic conceptualization and framing of nature as a whole, including the elusive character of any single, unified understanding of the

[48] Millán, *Schlegel and the Emergence,* 143.

absolute as such, would begin to disappear, as the focus of neo-romantic naturalism shifted to looking within the mathematical physics (and chemistry, and biology) increasingly associated with ostensibly understanding universal "nature" in the most profound mode of which we were capable as human knowers. I am thus focused here on those foundations for the later, even more literal and even less subtle forms of ontologizing using relations of necessity, a theorized whole of a Spinozist kind, and a new mythology of reason interpretation of the whole enterprise, which the German thinkers did, in my reading, indisputably build, and which were so diametrically opposed by Jacobi.

One of the key features of the early German romantic philosophy of the new "nature" which would prove so attractive to later thinkers, and particularly scientists, was not so much the content of their philosophy but its method, including its deliberately informal quality that broadened the avenues through which an individual could think or speculate upon nature directly and intuitively, while still retaining license to absolutize or infinitize. This departed from earlier (theistically or deistically infused) approaches to metaphysics and ontology and reflected the romantics' interest in pantheist, panentheist, and atheist reframings of reality along with their dissatisfaction with, and reaction to, other philosophers of their day. As we have noted, the romantics drew heavily from the then-popular Herderian reinvention of Spinoza in their philosophy. They also drew from other thinkers like Jacobi, Kant, and Fichte what they wished, while lambasting other aspects of their thought. The romantics were magpies who knit their approach together with imagination, most visible in the piercing poetry and ironic philosophical fragments of Novalis and Hölderlin, and the ironic philosophical writings and aesthetic critiques of Schlegel. Schelling was the most traditional philosopher of the four, but even he was untraditional in terms of the sheer volume and brimming creativity with which he wrote and continually reinvented his thinking, and the ease with which his system, especially in his *Naturphilosophie* years, claimed to make direct speculative contact with nature and know it as ultimately also [derived from] Nature.

In their efforts to know or represent nature as ultimately Nature, all four of these writers worked within ontological and epistemological frameworks (or, more loosely, largely intuition-based assumptions about nature and human thought) which were neither strictly realist nor strictly idealist, but both. There are of course a number of ways to define the distinction between idealism and realism. In this context, suffice it to say that roughly speaking, an idealist views ideas as somehow ultimate, while a realist views some other kind of "reality", often the physical universe, as ultimate, with the status of our ideas and knowledge less certain unless they are claimed to mesh somehow absolutely perfectly with this reality. For the idealist, world is a product of idea (or sometimes "mind"); for the realist, idea is a product of world, and it varies according to the thinker how we think that this production happens, how much distortion it leaves embedded in our ideas, how incomplete and partial they must be even in terms of their formal qualities, and so forth. It is also common in philosophical speech to use these designations adjectivally. To say a claim is idealist usually means that it is meant to be understood (or evaluated for truth or falsity) vis-à-vis some proposed realm or regime of the inter-relation of ideas. To say a claim is a realist claim, on the other hand, usually means simply that it is intended to hold true of or apply to reality itself (however this is conceived by the thinker in question) and that it should be tested or evaluated somehow "against" that reality, often via observation or experience.

Some idealists propose a truly independent realm of ideas which does not seem to connect to what we think of as physical reality, but usually, idealists connect the ideal realm to the real world we inhabit, either as this really experienced world's ultimate explanation and order and/or as the seat of its active controlling and directing arch-mind (for most of Western thought, until about this 1800 watershed, associated with some form of divine mind). Idealists tend to be rationalists epistemologically, i.e., assume that the functioning of [usually logical, theoretical] reason, without full reliance on observations per se (and sometimes even with the intentional casting-aside of all evidence derived from experience or the realm of the senses), can lead us to know and reconstruct this ideal realm and so progress to true knowledge of the highest kind of which we are capable. For the realist, again, our ideas (including any elaborate systems or "realms" of ideas we construct as human knowers) can have varying kinds of connections to what is posited as the real objective world beyond us. Realists usually (though not always) begin with the physical world we can observe and experience and then try to explain it; epistemologically speaking, realists tend to be empiricists because they think that the real world is apart from us and empirical data-gathering is the best way we have of continually interacting with it in order to progress in our knowledge of it.

Inserting the term "critical" before either idealist or realist usually signposts adherence to (or acknowledgment of) some element of Kant's revolutionary critique that had caused a ripple (or an earthquake) of deep unease over the status of secular reason itself around the time of the Controversy as previously noted above, effectively by exposing the flaws in both rationalist idealism and empiricist realism as pathways to truly objective or final kinds of truth. A critical idealist would usually want to integrate into their view some type of Kantian acknowledgment that seemingly perfect or non-falsifiable (transcendental, for example) arguments cannot really offer us ways of knowing about ideal things or metaphysical questions absolutely or with scientific-like certitude.[49] Likewise a critical realist would usually want to accept some version of Kant's caution that even scientific reason has a pre-templated way of approaching the external world using unavoidable categories we cannot change, and that we are cut off from knowing our physical objective reality as it is "in itself"[50]; we always get only our own picture, not a pristine God's-eye view.

Sometimes, as with the early German romantics, and arguably also with Spinoza, philosophical approaches to reality seek to calibrate their position to embrace both realism and idealism. These philosophies are often quite maximalist about either human epistemological access to ultimate reality using our powers of induction and deduction, or what the physical world divulges to us right out in the open (or with some straightforward observation), or, as in the romantic case, both. Usually the basic thematic claim pertaining to nature in such dual systems is something like "nature as a whole is really and objectively ideal." In the early German romantics' dual system, necessity played a key role, serving as the bridge between nature's core intelligibility and its objective reality. Schelling held that Spinoza does not begin his system with the Cartesian *cogito ergo sum*, but instead a new, "very different concept, quite independent of the I, the conception of a connotation of all possibility[51]". This implies that some set of "all possibility" defines the real world and, if you read Spinoza as a realist of some type, as the romantics and Herder did (cf. below), this set really "is," and is not solely mental (it is "quite independent of the I"); at the same time, it offers the thinker a handle of fundamental intelligibility, even a master-idea, by which to logically reason about nature as a whole and in this way implies an ideal realm (Nature's own mind, or the panentheistic God, as

[49] Cf. Robert Stern and Tony Cheng, 2023, Transcendental Arguments, in *The Stanford Encyclopedia of Philosophy*, eds. Edward N. Zalta & Uri Nodelman, and Kant, *Critique of Pure Reason*, 1781.

[50] Cf. Kant, *Critique of Pure Reason*, e.g., Book I/ Second Part/ First Div./Bk II/Ch. III.

[51] F. Schelling, 1975, On the possibility of a form for all philosophy, trans. Fritz Marti, *Metaphilosophy* 6, 15.

we will return to in what follows) containing, directing, or circumscribing nature's own unfolding. Not everything is possible; things are only knowable as possible if there are also impossible things. Impossible things must violate rules established by our understanding of logical or mathematical necessity, since there is to be no other standard for what is possible versus impossible, none other than necessity or its absence, in this newly depersonalized totality. Necessity as a principle and as a characteristic is the [bare] criterion for existence. This idea here attributed by Schelling to Spinoza, of the world as [only] "all possibility", is arguably one of the core aspects of nature all four of our thinkers adopted and sought to infinitize, elevate, and romanticize in their philosophy. It would also turn out to be a conception of nature which would have enormous ramifications in later ages of the West.

Interestingly this conviction of fundamental necessity-based intelligibility also helped the romantics think of even chaos in nature as "pre-form," holding it just within the scope of reason.[52] The German romantic worldview, as noted above, often imagined infinitude-finitude connections as elevating to finite human life, but these connections also destabilized or threatened finite human reason, since to actually think what its link to the infinite meant for the finite realm was to imagine the integration (in loosely the mathematical sense), up into the unity, of all the infinitesimals of finite experience and phenomena, and thus a type of death or sublimation of the individual and the world. Contemplating this, or any conversion of the

[52] "*Chaos*," Schelling writes, "is the fundamental intuition of the sublime;" we had to see natural things as "symbols of the infinite" partly through a "fundamental intuition of chaos" which was part of "the vision or intuition of the absolute." This also worked into his fusion of philosophy of art with his philosophy of nature; in a sense, to see chaos in nature could be overcome by making it into the just-tolerable aesthetic experience of the sublime, in other words by learning to see in nature what the romantics held we could see in [particularly tragic] art, even if the sublime pressed art to its very limits. In his *Philosophy of Art*, he calls primal chaos the "inner essence of the absolute" which is both unity and "primal chaos itself"; again, this was not raw chaos, not Other to the one unity but rather a dizzying reveal of it: it was a locus of the "identity of absolute form with formlessness"; not a "*mere* negation of form, but rather formlessness within the highest and absolute form, and, in a reverse fashion, absolute form within formlessness." It was, again, nature's creative unconscious, to parallel man's (cf. Hölderlin on this concept as well, below). In Theisen's view, Schelling was able to give "chaos a positive value as a concept of form." Cf. Friedrich W. J. Schelling, 1989, *Philosophy of Art*, ed. and trans. Douglas W. Stott, Minneapolis: University of Minnesota Press, 88 - 89 (no. 3.65); Bianca Theisen, 2003, "χα Absolute Chaos: The early Romantic poetics of complex form", trans. K. Barry, *SiR*, 42, 306.

finite into the infinite, was akin to contemplating chaos, and this was also the chaos the romantics at times glimpsed in nature's terrible power and austere indifference to human affairs. Still, again, even chaos had to be somehow amenable to or encompassed by reason, due to the overriding fact of the unity of absolutely all (and its groundedness in necessity). Hölderlin coined a neologism for the chaotic in nature, the "aorgic", which he meant as a negation of "organism-like". The aorgic stood in a pair-like or dialectic relation with the organism-like or the formed. All of nature, before man's transforming presence, remains close to or absolutely identified with the aorgic. Man is the ultimate in form, or organism, because of his powers of creativity and conscious thought. Part of man's highest calling is to confront this deepest aorgic aspect of nature, to absorb even this chaos into his own "self-activity and art and reflection", through an art which is also transformative of nature; nature becomes "more organic through man, who forms and cultivates, through the creative drive and creative powers in general". At the same time, by integrating even this chaotic aspect of nature into himself, man "has become more aorgic, more universal, more infinite[53]".

Schelling's approach to keeping chaos [just] within the frame of reason is less embodied in its orientation, but not unrelated to Hölderlin's. Schelling writes:

> After unsuccessful attempts to exhaust the chaos of the phenomena in nature and in history by means of the understanding, ordinary perception or knowledge resolves to take "the incomprehensibility itself," as Schiller says, "as a principle of judgment." This appears to be the first step toward philosophy, or at least toward an aesthetic view of the world.[54]

Even if we have to assign the chaos incomprehensibility as its unifying principle, it is still rendered, in this way, intelligible. Reason holds in the face of the infinite or the chaotic, saved by its capacity to decide something is only thus-far incomprehensible and thereby keep it, if barely, within the frame of necessity and not exclude it as impossible. Foreshadowing neo-romantic claims to a "Copernican" vantage point (and here opposing Kant) in a way which will come to distinguish later scientific views of nature, Schelling even tries to make a virtue of chaos's opacity-to-us:

[53] Hölderlin, 2009, Ground of the Empedocles, in *Friedrich Hölderlin: Essays and Letters*, ed. and trans. J. Adler, and C. Louth, London: Penguin, 261–262.

[54] Schelling, *Philosophy of Art*, 88 - 89 (no. 3.65).

"Understanding," he writes, "can recognize the world as the true symbol of reason itself only within such an unbounded condition, one appearing to common understanding as lawlessness[55]". Lawlessness and total freedom, which at first threaten our ideal world when we turn to the seemingly unlimited chaos of the all-that-is or nature in its raw infinity, resolve into integral, vital parts of the set of all possibles: they provide the very framework which allows us to make sense of law and necessity themselves. The infinite set of all possibles is, effectively, Nature: considered as infinite, it appears chaotic, but understood as nature, as the set of all possibles enframed by something which allows us to perceive how law and necessity differ from lawlessness and total freedom and thus to understand something intelligible about its constitutive core, it is intelligible. Arguably the only way to claim this type of insight into nature is via an atypically total combination of idealism with realism, and a fusing of human epistemological limits and boundaries with nature's own ostensibly real lineaments and characteristics.

In another example (already touched upon above) of how the romantics' version of Nature-nature fused realism with idealism, because of its Spinozistic monistic unity with (or containment of) absolutely all that existed, romantic philosophy found a new starting point for thought, one which updated the too-rationalistic Cartesian *cogito* with a turn to thinking in and as nature (idealized as sheerly the unified totality of all things). Schelling counted as his bedrock one " 'real ... true knowledge[56]' " the fact that we are [all] directly *of* the infinite absolute. Hölderlin rested his truths on a single intellectual intuition of "unity with all that lives", the knowledge that the "objective as such, and also the subjective as such, are only a condition of the primally-one[57]". Schlegel held that despite the "infinite separation between consciousness and the Absolute", as Bullock puts it,[58] philosophy nonetheless takes place in the endless effort of the "I" to reach or conceive of just this absolute itself. Nature was amenable to being thought and known from this type of departure point because nature and mind were not fundamentally separated. In Richards' view, Schelling's whole philosophy held nature "the self's other kingdom, where

[55] Schelling, *Philosophy of Art*, 88 - 89 (no. 3.65). , pp. 88–89.

[56] Bowie, *Schelling and Modern*, 61, citing Schelling [I/6 140] in his *System of the Whole of Philosophy and of Naturphilosophie in Particular (1804)*, or the "Würzburg System" as Bowie refers to it.

[57] Hölderlin, 2009, Untitled fragment, in *Essays and Letters*, 303–304.

[58] Marcus Paul Bullock, 1980, *Eclipse of the Sun – Friedrich Schlegel, Walter Benjamin, from new mythology to the theory of new history*, ProQuest, 212–213.

the forces were familiar and the mind would meet its double, most beautifully displayed[59]". Beiser notes that what characterized the "ideal" for the romantics was ***not*** "the mental, subjective, or conscious" but the mind-like in the sense of being "rational, archetypal, or intelligible[60]", traits which could then ostensibly be uncovered in the natural world around us as well as in our own thought. As Hamilton Grant puts it, "whereas analysis and synthesis were powers of the understanding for Kant, for Schelling, they are powers of nature[61]". (This is also implicit in Schelling's diagnosis of the *Grundirrtum* of traditional philosophy noted above.)

The early German romantics did not shy away from embracing the novelty, what I will shorthand per below as the quadraticity, of their "nature". They deliberately used new names for their philosophical approach, such as Novalis's "magical idealism" (sometimes "magical realism") wherein, as Wanning notes, we can see Novalis's notion that metaphysics itself can also become a kind of "poetry[62]". Beiser refers to their approach as "absolute idealism.[63]"

Folded into this combined idealism-realism, and related to it, again, the romantics considered the human mind's epistemological boundaries to reflect nature itself, not any insurmountable human limitations; as our knowledge grew in ages to come, we would be able to explore ever more of nature. Any sound way of forming knowledge of nature which our intuition approved, could be trusted, whether artistic or scientific-mathematical. In the more idealist sense, since mind was one with nature, romantic thought could also legitimately "make" new layers of reality by actively transforming the world, as in the Hölderlin example of man as the conduit for making nature more formed and organic noted above, or in Novalis' view of the proper function of the romanticizing or elevating power of the human mind (cf. below on their new mythology of reason project). The distinction in traditional philosophy between epistemological concerns (concerns about the correct methods and forms for true knowledge) and ontological concerns (concerns about what can be said to really exist in the world), always arguably tenuous, in the romantics becomes ever more

[59] Richards, *Romantic Conception*, 116.

[60] Beiser, *German Idealism*, 353.

[61] Iain Hamilton Grant, 2010, Introduction to Schelling's *On the World Soul*, in *Collapse* VI, ed. R. Mackay, Falmouth: Urbanomic, 61.

[62] Berbeli Wanning, 2017, Poet and philosopher: Novalis and Schelling on nature and matter, in *Ecological Thought in German Literature and Culture*, ed. Gabriele Dürbeck, et al., Lanham: Lexington Books, 47.

[63] Cf. Beiser, *German idealism*, e.g. 355; and Beiser, The Paradox, 217–237.

blurred. Nassar more conservatively on this point notes that these thinkers were prone to combine epistemological and ontological concerns into one mostly due to their great interest in understanding how, and why, knowing *did* at times cleave off from being and diverge from it; for Nassar these romantics saw both the unity and distinction between being and knowing.[64] Others have also read them as slightly more critical, or hesitant to give full ontological weight to anything they claimed to know about nature as such or as a unity[65] (other than that the unity itself was present and real). But I would argue that overall, their ambitions to conceive of nature as also Nature, and as encompassing absolutely all, were mind-centered and equipped them to both uncover and create, in their view, true knowledge of this Nature-nature as the real world. In my reading of these thinkers, the very operation that Jacobi was arguing needed to only be carried out in a negative sense (i.e., to ensure that we did *not* think our own epistemological limits actually curtailed the real, or mistake for the real one our "world" we could perfectly conceptualize as a necessity-bound entity) was transformed in these romantics' hands into a positive pathway for revealing nature as it really-ideally was and as it could be further transformed into being.

Thielke observes that Schelling, overall, was "attempting the admittedly difficult task of defending monism" in light of Kantian critical philosophy.[66] In addition to conceiving of these thinkers as striving to occupy the standpoint of production rather than reflection (per above), we can also bookmark this tension-filled notion of a critical monism, also present in Bowie's reading of Schelling and to some extent Nassar's readings of

[64] Nassar, *Absolute,* cf., e.g., 259–260.

[65] Frank and Bowie, and to some extent Larmore, treat of Schelling on this front apart from his three colleagues. Cf. Bowie, Friedrich Schelling; or Bowie, *Schelling and Modern*. Frank reads Hölderlin, Novalis and Schlegel (but not Schelling) as more skeptical in terms of their epistemology, and as more averse to making direct statements about the one unity itself. Frank reads Schelling as far more willing to try to "take hold" of the absolute using reason, attributing to him a "combination of … monism and … realism" also somewhat akin to that of Schlegel and Novalis, cf. Frank, *Philosophical Foundations,* 24, 55; 56 for quotation. Cf. also Frank, 2016, 'The Difficult Step into Actuality': On the Makings of an Early Romantic Realism, *Comparative and Continental Philosophy*, 8: 199.

For more on some of the thinkers' ontological reticence cf. also Frank, 'The Difficult Step', 212; Frank, *Philosophical Foundations,* 28; Charles Larmore, 2017, Hölderlin and Novalis, in *The Cambridge Companion to German Idealism,* Ed. Karl Ameriks, Cambridge: Cambridge University Press, 205–226.

[66] Thielke, Who's Who, 400. Bowie claims this as well, cf. Bowie, *Schelling and Modern,* 5.

Schlegel,[67] as a useful one, to which we will return later. But again, if the early German romantics embodied a special kind of self-conscious turn "toward nature" and away from Kant and Fichte,[68] while also seeing themselves as remaining bound by some of Kant's critical insights,[69] overall their turn to nature overrode Kantian caution, and brought a new level of philosophico-poetic-mathematical assertiveness to thinking nature directly. Rather than thinking of them as establishing a fixed set of beliefs about nature, in our Controversy context it is more important to recognize how their "quadratic" (my term) nature was forged methodologically, so we are equipped to recognize it when it returns in later Controversy iterations. Whether intentionally or not, the romantic legacy would be of a particular type of "nature" wide open for direct investigation and experience by the thinking "I" and always promising to reveal its unity with, and convertability into, this "I", but also linked with infinitude and the absolute, however problematically for the finite mind and however much it was just-able to keep such an overwhelming ontology rationalizable. Their monism meant there was only the one unified continuum, and nothing else. This "nature" combined idealism with realism, holding nature to be absolutely and really ideal; and its ontology or catalogue of being meshed with what it was possible for the human mind to both uncover and make within nature, using artistic creativity and science harnessed to the intuition. I use the term quadratic to shorthand this nature not only because of the fourfold it entails (ideal-real and epistemologically accessible or mappable-ontologically significant) but also because the romantics were convinced that human knowledge needed to be consciously exponentialized or intensified, often discussing how to raise the idealist's insight into nature by the power of making it more real, and how to raise the power of the realist's insight by making it more ideal. Novalis writes:

[67] On Schlegel, cf. Nassar, *Absolute*, 157. On Schelling, cf. Bowie, *Schelling and Modern*, 34, 31.

[68] Cf. e.g. Dalia Nassar, 2012, Spinoza in Schelling's early conception of intellectual intuition, in *Spinoza and German Idealism*, 155, regarding Schelling's turn from Fichte.

[69] Again, for this point, cf., e.g., Nassar on Schlegel's maintaining a "'critical', or post-Kantian, philosophical comportment without, however, accepting the Kantian or Fichtean limits of knowledge or their understanding of the relation between mind and reality," Nassar, *Absolute*, 157; or cf. Bowie, who notes that Schelling at times would speak about "nature in itself, albeit whilst acknowledging certain of Kant's strictures" insofar as he foregoes simple "*dogmatic* assertions about nature in itself", trying to thread the needle by hewing to theories about subjectivity in nature, cf. Bowie, *Schelling and Modern*, 34, 31 (emphasis added).

> The world must be made Romantic. … To make Romantic is nothing but a qualitative raising to a higher power. … By endowing the commonplace with a higher meaning, the ordinary with mysterious respect, the known with the dignity of the unknown, the finite with the appearance of the infinite, I am making it Romantic. The operation for the higher, unknown, mystical, infinite is the converse—this undergoes a logarithmic change through this connection—it takes on an ordinary form of expression. Romantic philosophy. Lingua romana. Raising and lowering by turns.[70]

The rendering of nature such a quadratic, however subtle or aspirational in the romantics' own treatment it at times seemed to be, left a pathway later interpreted as leading to ultimate knowledge of the single Nature-nature. As the romantics' irony and even despair which at times complicated their efforts, and ultimately (in part) led them to abandon them, fell away from their core image of nature which lived on, and their approach was reinvented in ever more scientific and less philosophical terms, their quadratic nature would be changed into a version of itself perhaps not possible for the romantics themselves to fully anticipate, but arguably quite accurately anticipated by Jacobi.

A central element of their approach which both allowed their quadratic to persist so durably, and allowed it to later take on [even more] deeply science-intertwined forms, was, again, the concept of necessity. While we have already discussed necessity in part in the foregoing, these thinkers' views of necessity in—or even as a proxy for—nature bears briefly examining further here, before we proceed to an overview of their "new mythology of reason" project.

The New Nature of Necessity: Romanticizing Spinoza's Pillar

While there is not room here to address Spinoza's original positions in detail, it may still be useful to preface this section by touching upon a few aspects of how necessity characterized [all of] being in his philosophy. As noted in Chap. 1, Spinoza was a theist, or (since his theism is debated, and we cannot do the question justice here) at the very least consciously writing as an inheritor of medieval monotheistic traditional proofs of the existence of God, proofs quite different in form and function from proofs undertaken in purely secular contexts, not least because of the way the theological type relied upon a[n often premised, axiomatic] concept of

[70] Novalis, 1997, Logological Fragments I (1798), no. 66, in Philosophical Writings, ed. and trans. Margaret M. Stoljar, Albany: SUNY Press, 60.

God. Spinoza's concept of God or Nature as the single necessary, because single "self-caused", existent—"By that which is self-caused," he writes, "I mean that of which the essence involves existence, or that of which the nature is only conceivable as existent" (*Ethics*, I/Def./I)—should be understood in this context. Furthermore, as Zylstra notes, Spinoza's philosophy still "preserves the core conception of efficient causality as the bestowing of existence on something" which is a "conception [which] traces back to Avicenna and was widely accepted into the seventeenth century.[71]" This notion of an efficient cause of something as that which bestowed existence upon it had been part of the medieval theological account of the world, wherein a necessary, self-caused, transcendent deity bestowed existence on, or caused, a contingent (not necessary) physical world. This pattern of distinguishing between necessary and contingent beings, and relations, had also then been repeated at multiple smaller levels within the medieval world-picture, including in the era's neo-Aristotelian fourfold classification of causation which distinguished among, in effect, various kinds of causal necessity versus causal accident or contingency. But Spinoza took away contingent things and discarded the Aristotelian causal classification system. In Spinoza being belongs only to the self-caused real entity, God or Nature, and the only things which are, are in this same way. The world becomes as necessary as God, because it is no longer separate from God. Causing something was still associated with bestowing existence on it, but Spinoza made this a singular case, flattening traditional arguments for the existence of God into an argument for the existence of God and world as one (anticipating Jacobi's later observation that naturalists must always prove their "nature" exists just as much as they must prove their "God" does not, per Chap. 1). Spinoza effectively tried to prove first the existence of some one thing and then show how no other thing besides this one thing could be, or be real. This singular case was not the same as the traditional Avicennan case where existence is transferable (from God to another kind of being) and necessity is invoked as part of meeting the threshold for a satisfactory, logical, causal explanation of world and God both existing [differently]. In Spinoza, logical reason is used to turn the ontological screw much tighter. For Spinoza, "God must be the sole cause, inasmuch as to him alone does existence appertain" (*Ethics*, I/Prop. XXIV, Cor.), and "although each particular thing be conditioned by another particular thing to exist in a given way, yet the force

[71] Cf. Stephen Zylstra, 2023, Spinoza, Emanation, and Formal Causation, *Journal of the History of Philosophy*, 61: 605.

whereby each particular thing perseveres in existing follows from the eternal necessity of God's nature" (*Ethics*, II/Post./Prop. XLV/Note). The way we understand causes of things existing/not existing should at most be understood as speaking about causing different conditions or configurations of their existence, not their real or actual existence as such. There was to be no more transferable, mutably present or not-present existence in the absolute sense; and there was to be no more real contingency at the substantial level. Everything that was, was God, and exhibited the necessity inhering in a self-caused (and thus absolutely necessary, because unimaginable as not-present) substance. Necessity was in all things; it was no longer the sole purview or power of a person-like divine ruler. Spinoza sunk necessity "down" into the plane of the world itself, or lifted the world "up" to the plane of necessity.

What then happened later, in the romantic context, was a revamping of Spinozan necessity into a new romantic-Spinozistic version. From Spinoza's necessity, uniformly characterizing the only substance that was, defining a new form of monism and assuring us of its reality (especially via proving the non-separation of God from world), the romantics crafted a new version, one which took this monism, this unity and oneness of God and Nature (and mind), as already established, and used necessity in a more active sense. Necessity was seen as being able to do more than just show us why there must be the one unity: for the romantics, tracing necessity relations could also help us know this monist whole directly, via mathematics and science; via rational, true philosophy of nature-oriented, speculation; and via art. This meant necessity relations provided a way, in all three realms (and all three realms were understood to be fundamentally one, as we have been discussing), to reconstruct the universe and trace its real lineaments, to see the world through Nature's eyes as necessity manifesting necessarily.

If we think of the notion of necessity as the encapsulation of a claim that something *cannot be otherwise*, there are two basic kinds of necessity claim (and resultingly two kinds of necessity relation) any rational system might try to build a "world" upon: 1) that "X cannot be *otherwise*" referring only to some logical or mathematical outcome set up and tested within a given context or system and ultimately only referring back to the system's pre-designated rules and premises; or 2) that "X cannot *be*, otherwise" invoking some condition for something to really exist where the terms are ostensibly absolute and historical, really speaking of the presence or absence of something in its entirety. For Jacobi, 2) is the fiction-generating tool of naturalistic speculation or rational ontology of any kind, whose "world" constructors deceive themselves that they are exceeding the mere terms of 1). Both kinds of necessity-claim reduce to definitional circularity, for Jacobi, who like Hamann is nonplussed by invoking "necessity" as though it meant we somehow broke out of these circles, instead of only marking some kind of certainty we value, which means, some certainty whose definitional solidity we ourselves have created (cf. Chap. 1). For Jacobi, as we have seen, questions of real being and existence must be reserved for metarational perception and reason, not surrendered to such a theoretical shadow-play version of asking and answering what makes things be, this false dialogue (in the romantics, often a particular type of romantic dialectics) which blocks and renders nothing the dialogue with the actual world the metarational subject must be allowed to preserve and sustain. Spinoza, for his part, employed both kinds of necessity claims to ontologize monistically, but in a deliberately self-referential way, in effect showing that if we are to ontologize using these types of necessity claims we can at most do so singly, for one case; all roads in Spinoza lead to a single, perfectly existent, perfectly necessary, God or Nature. For the Herderians and the romantics, again, while the notion of an overarching unity of all arising from Spinoza's monism could be retained, his monistic flattening of the notion of necessity could not quite be left as it was. They held that interacting with nature, we could see or uncover necessities of the 1) or 2) kind which promised to be the lineaments of something like Nature's own necessity unfolding itself [in and even as nature]. As touched on earlier they also added a third, specifically romantic, type of necessity relation to the first two: an artistic type of necessity invoking the intuitive, creative, aesthetic perception (or near-blinding glimpse) of the absolute, leading to another world-building necessity relation of a mold akin to 3) that X cannot be otherwise *than as this work of art makes or reveals it to be.*[72]

Without Spinoza's disciplined deconstruction of it to hold it in abeyance, the West's long monotheistic habit of associating rationally derived statements of necessity with actual ontological effects returns, but now in a system perceived as pre-saturated with or even solely "made of" necessity alone. In direct contrast with Spinoza's own aims, the net effect of the romantic universe is to lower (rather than raise, as Spinoza seemed intent on doing) the threshold for perceiving a necessity relation as truly

[72] Cf. also Eichner on this point, who notes that with the new "Romantic view of poetry as the fountainhead of all higher truths ... ordinary language" could no longer suffice; the "higher truths that are revealed through the irrational or suprarational powers of the imagination" relied on "symbol," and poetry now had to be symbolic and carry these truths; there was no more room for the idea of art as mere imitation. Eichner, Rise of Modern Science, 19.

ontological or revealing of something true about real being. As long as one of the three kinds of necessity claim could be discerned or created, via art or thought (and then later, also experimentally tested and proven), the romantics' world really was, and was necessarily, leaving no way out of it and nothing outside of it. When Spinoza's position that "God is the sole free cause. For God alone exists by the sole necessity of his nature … and acts by the sole necessity of his own nature" (*Ethics,* I/Prop. XVII/Coroll. II) begins to become confusedly transmuted into license to take any necessity statement as evidence or documentation of the activity of the one real, God or Nature, and how it really makes things according to necessity such that they cannot be, otherwise, then mathematics and science, rational speculative naturalism, and art become the sole means humans possess of accessing real truth. The metarational awareness or frame is no more; it is banished. In Jacobi's view, the human loses the world. Rather than inhabiting the world, human minds must continually [expose and] make it through the lens of the three kinds of necessity relations alone. In the romantic era, all the work Spinoza did to ensure all roads led to the one substance, all his labor to build out the reasons why "Nothing in the universe is contingent, but all things are conditioned to exist and operate in a particular manner by the necessity of the divine nature" (*Ethics,* I/ Prop. XXIX), becomes increasingly interpreted as license to absolutize any kind of necessity statement science, rational speculation, or art can generate. Even the obviously creative and inventive dimensions to such a process were not taken as faults, but as further merits of the new way of thinking-making universal "nature." All Spinoza's efforts to move all of reality into one kind of plane, rather than being understood as a challenge to elevate our understanding to try to think in terms of this single Ur-substance, are turned on their head. The tool he used to try to show reality as more profound, necessity (and our capacity in general to understand a concept like necessity objectively, and follow where it leads; our capacity to guide our own rational speculation according to logical processes, rather than capriciousness or slavish adherence to inherited revelation), now begins arguably to be used to move reality in the opposite direction, toward being seen as an entity of absolute shallowness (though this will take centuries, as we will return to below). The romantic universe "made" of knowable necessity-relations takes the Spinozan set of all possibles, which Spinoza seemingly used to frame and ground his giant endeavor to know all of reality as one kind of thing (cf. below), and begins to tame it into a gloss on romantic theoretical-speculative-artistic reason's capacity to, at least in principle, frame all of what infinite nature itself really is, and do so "without mediation", again in contradistinction, arguably, to Spinoza's insistence on God or Nature—reality—as possessed of an infinite number of attributes (*Ethics*, I/Def. VI), of which thought and extension in their entirety were only two (*Ethics*, II/Props. I - II). To understand only one or two attributes' core characteristics, out of an infinite number of attributes, even were this achieved, must still leave effectively all of the God-substance untouched by human knowledge, even if it offers us the consolation prize of monism itself as the only all-encompassing understanding of being of which we are capable given our own finitude. But with each successive wave of romantic interpretation of Spinoza emerging out of each Controversy iteration, the distinction would erode further between, on the one hand, using necessity-relations to define or reconstruct some type of "world" we understand is only an infinitesimal part of the full reality in which we find ourselves; and on the other hand, claiming to, via necessity relations, in principle be positioned to completely know God or Nature, the only substance, the real itself, that which Spinoza himself had originally emphasized must be understood as "absolutely infinite being ... consisting in infinite attributes, each of which expresses a certain eternal and infinite essence" (Ethics, I/Prop. X, Note).

Spinoza's own monist system, where the thinking "I" claims to perceive how everything already "is" and "is" necessarily, was read by Jacobi as an ingenious yet ultimately failed attempt to offer a realist account of the world[73]. In the Controversy, Jacobi was concerned that Spinozan necessity, misread as the kind of manipulable and knowable necessity useful for "weaving" the new naturalistic world made of "blind forces," raises the specter of a "weaving of weaving" world from which metarational awareness must be completely removed (because it must always "recoil in horror" from it, as discussed in Chap. 1). Jacobi, unlike the romantics, saw that Spinoza himself only knit together the whole of being in one single fully necessary substance which was effectively nondynamic and was quite different from the world claimed by the new naturalists; indeed, in Sandkaulen's view this was one of Jacobi's main contentions with Spinoza's system itself – that it can only address eternal time, and not finite existence.[74] But Jacobi also saw that Spinoza had left a tantalizing precedent for which he *could* be faulted in that he offered an object lesson, for those

[73] Re Jacobi's reading of Spinoza as realist cf. Sandkaulen, Jacobi and Spinoza, 45.
[74] Sandkaulen, Jacobi and Spinoza, 46.

who followed him, in using theoretical reason alone to muscle through to an attempt at a full account of reality, full stop. Sandkaulen holds that for Jacobi, "[t]aken as a self-contained metaphysics of immanence, Spinoza's philosophy cannot be refuted" but that Jacobi undertook to *contradict* it [Spinoza's philosophy] in full, because of what we know and experience of the real world in our actual experience; "[i]n Jacobi's assessment, the exclusion of *causa finalis*, along with the paradoxical alienation of time, is systematically inevitable but, from a factual point of view, unacceptable;" at the same time, "our original experience of action," as noted above something Jacobi thought must always animate reason, "must be at work in the depths of Spinoza's metaphysics," but simply be "disguised.[75]"

I would argue that in the context of the Controversy, and particularly in my environmental reading of its importance, the most important aspect of Jacobi's "contradiction" of Spinozism was his rejection of the "weaving of weaving" view of the world threatening to arise out of the romantics' invention of more manipulable versions of necessity-existence links than Spinoza's austere own.[76] Using all three kinds of necessity-statements to claim to be able to ontologize "nature" as a whole would be the lasting legacy of the Controversy, and this was part of what Jacobi was trying to fend off, due to the nihilism such a practice represented, in his view (cf. Chap. 1, and below). He would fail, and exuberant romantic necessitarian images of nature would be left like a minefield for Western thought thenceforward. From the Jacobian metarational frame, romantic use of the two (or three) kinds of necessity claim to ostensibly viably "weave" what nature really is, and the more disciplined notion in Spinoza himself that weighing and reconstructing ["just"] one type of [master] necessity-relation using theoretical reason could "knit" a perfect and really realist monism, were simply different facets of the same worthless false gem. Necessity is just an ordinary concept of the theoretical, circular, self-referential kind of reason, and this kind of reason is not suited for looking "out" at the world as it really exists. Necessity is not a symbol of contact having been made with something like the real itself, or "nature"; it is merely a signpost of human certainty (cf. below). Necessity relations do not bestow real existence or carry ontic importance or reference (as Kant

[75] Sandkaulen, Jacobi and Spinoza, 47–48. On 48 Sandkaulen claims that for Jacobi, "Spinoza's metaphysics is based on the foundation of nothing other than a philosophy of action", but "[i]n the interest of systematic justification, this foundation is disguised."

[76] Again, for further on the theological inheritance at work in Spinoza's understanding of causation (and necessity) cf. Zylstra, Spinoza, Emanation, and Formal; or Douglas, Was Spinoza.

also partly implied when he argued against using "existence" as a meaningful predicate for more straightforwardly linguistic reasons[77]).

The minefield romantic Spinozism left behind becomes even more pronounced when the romantics (and later, neo-romantic scientists) try to increasingly concretize Spinoza's concept of nature as the sum total of necessity relations permitted by a [distortedly] naturalized version of what, in Schelling's view (as noted earlier), Spinoza had set up as his template for truth, his "principle of all form", his chosen a priori that established the pattern and the boundaries for his system: Spinoza's new, "very different concept, quite independent of the I, the conception of a connotation of all possibility."[78] It seems Schelling is roughly correct in this observation of what undergirded Spinoza's monist system, even just judging from, again, the number of times Spinoza's *Ethics* designates the only alternative to necessarily caused as "impossible", e.g., "it must follow, either that a triangle necessarily exists, or that it is impossible that it should exist" (Part I/ Prop XI/ Proof)."[79] For Spinoza to create his complex, holist account[80] of the one existing substance, he seemingly still required a limiting set of all possibility, or all potentially true expressions of the necessary nature of God at least as expressible in our worlds of thought and extension, which were in turn, again, only two of God's infinite attributes (cf. e.g. *Ethics*, II/Props I-II). Part of Spinoza's legacy will be that outside of his theological context (in which it seemed to be his primary intention, per above, to use necessity as a lever to arrive at monism and pan(en)theism, not as a lever to absolutize necessity as it appears in statements of the 1) - 3) type just discussed), his seeming axiomatized presumption of something like necessity (rather than not-necessity) characterizing his single real substance in terms of [something like] its extent, or range, or degrees of freedom, as well as characterizing how or why it could be known as existent with certainty, will lead to romantic thinkers, and especially their later 20th- and 21st-century successors, transmorphing this imagined set of all possibles into something both mathematizable and able to stand for God or Nature completely or abso-

[77] Or perhaps because it complicated the analytic-synthetic distinction. Cf., e.g., Kant, Critique of Pure Reason, Book I/Second Part/ Second Div./Book II/Ch. I.

[78] Schelling, On the possibility, 15.

[79] As noted above, for quotations from *The Ethics* see B. Spinoza, 2009 [1677], The Project Gutenberg E-Book of The Ethics, https://www.gutenberg.org/files/3800/3800-h/3800-h.htm

[80] Sandkaulen emphasizes that Jacobi correctly saw Spinoza as undertaking a holist project, cf., e.g., Sandkaulen, Jacobi and Spinoza, 45.

lutely, and not only for one of an infinite number of attributes of God or Nature. The romanticizing of Spinozan Nature will also be marked by these thinkers' adaptation of another a priori which Schelling himself acknowledges must inform his philosophical efforts to think nature directly, and which sounds out a kind of cosmetic echo of the original Spinozan assignation of absolute ontological heft to the only substance that strictly speaking could be said (or rationally demonstrated) to exist. Schelling writes that "The philosophy of nature does not have to explain the productive power of Nature, for if it does not posit this as originally in Nature it will never bring it into Nature[81]". This helps complete the dubious romantic conversion of Spinoza's theological idea of the One into the transcendental condition for a scientific "ontology" of necessity encompassing the many.

From a Jacobian perspective, of course, before and beneath all these framing assumptions about Spinozistic [Nature or] nature, and how understanding necessity relations allows us to infinitely reconstruct or make it, what finally and in truth actually grounds the new naturalism is two acts of destruction, one directed against the external world and our ability to encounter it as a Thou, and one directed against a part of the subject itself. The belief that there is an entity like the set of all possibles which is both really true (and ontologically descriptive) of the world and also amenable to theoretical logical reason, is also not justifiable, for Jacobi.

In the romantics' dramatic conversion of Spinozan ontological, substance-seated, absolute necessity into something which also inhered in discoverable necessity relations of the 1) - 3) type, we can discern more of the special stakes for nature in the Controversy. *Pace* scholars who want to emphasize the romantics' Kantian caution, and Schelling's and others' emphases on how their only real, true knowledge was minimal, these thinkers still departed radically from Jacobi in their willingness to "uncover" such intense interplay between the theoretical mind and the external world, and grant such realist status to their idealized version of Nature-nature using necessity.

The romantics were willing to acknowledge that human ideas could never properly or directly grasp the absolute – acknowledge that their thinking of nature was a dyadic, oscillating construct entailing endless striving – but they held we could know there was no alternative on offer, just as committedly as Jacobi held their system was nonsensical because it barred the actually external world and its actual Otherness to the thinking

[81] Schelling, *First Outline*, 205–206 (nos. 289–290).

"I". For these romantics, even the never-directly-graspable absolute was part of their monist reality, and even human reason, guided and aided by the aesthetic sense and the creative imagination, was linked to it; the absolute, and it alone, was the final goal (as Nassar also points out)[82] of all our grasping and striving, in reason and in art, and in their higher forms, reasoned art or creative reasoning. There was thus in the end still only a single track of truth which was viable – the track which led toward the absolute itself and reflected our intuition and knowledge of its reality – no matter if our ability to travel this track must always remain partial.[83] Regardless of how Jacobian the romantics might sound, or how romantic Jacobi might sound, at various points *after* they choose differently on Jacobi's either/or, on the key choice itself, they are opposed.

Naturalist monism based on deploying the notion of necessity in a logical-scientific (or rational-speculative, or even artistic) manner to make a "nature" which is intelligible, systematized, and explicable is a powerful approach to the outside world because once its framework is accepted, it is impossible to find valid truth-claims within this framework which could fundamentally challenge or test it qua approach to reality. Errors or gaps according to the system can be corrected or filled, but nature will never be "found" to be unintelligible, or not a logically bound unity. Jacobi was keenly aware of this hermetically sealed quality of naturalism; this was why he essentially approached Spinozist naturalism, again, at a kind of perpendicular, opposing it as nihilism for its entire form and also for the whole self-stunting, reality-removing act he held it always required on the part of the subject. Since once the system "begins," it is in a sense too late to adequately critique or oppose it (or find footing to wholly "contradict" it, in Sandkaulen's term, per above), the anteroom of romantic naturalism is where Jacobi seeks to make his perpendicular intersect, and where the Controversy's center of gravity lies; this is why it is a Controversy, and not, at root, an internecine academic philosophical dispute between systems. Jacobi tries to argue against what has to first take place in the subject so that it can position itself not only to systematize but to believe its system is really true of the world. He faults what goes on in this anteroom, as we have seen, as involving the destruction of our recognition of the external world (and God), and the destruction of the fuller part of our reason.

[82] Nassar's reading, again, seeks to emphasize both the romantics' critical awareness, *and* their willingness to depart from Kant and hold reaching an idea (however imperfect) of nature as a whole as their goal. Cf. Nassar, *Absolute*, e.g., 5; 294.

[83] Nassar, *Absolute*, cf., e.g., 259–260.

What Halmi called the romantics' search for a combining of the subjective and the objective in a new organicist paradigm, backed by a "universal and inherently meaningful symbolism[84]" would soon prove irresistible, again, to the modern sciences in particular, as we will see in the next few chapters. Novalis writes, "The universe is the absolute subject, or the totality of all predicates. Both its immeasurable and measurable *structure* are based on this fact, for only in this manner does the totality of all predicates become a possibility[85]". The possibility of the totality of all [viable] predicates; the connotation of "all possibility"; mathematics and logic as universalized symbolism: romantic "nature" arguably rests ultimately on arbitrary statements of belief in necessity as an intelligible characteristic of the absolute itself, as the primary essence of God or Nature. Yet romantic "nature" would nevertheless soon be well on its way to becoming the West's main framework for understanding the real, and ostensibly sounding its complete ontological depth.

The German Romantic New Mythology of Reason

A third highly important and recurring (in both later Controversy iterations) element of these early German romantics' approach to nature was their hope that their innovative naturalist monism could lay the groundwork for a future "new mythology of reason". This idea was first broached in the "Oldest System Programme of German Idealism" written in 1796 by Hölderlin and Schelling (and Hegel).[86,87] The authors state that a new

[84] Nicholas Halmi, 2007, *The Genealogy of the Romantic Symbol*, Oxford: Oxford University Press, 63.

[85] Novalis, *Notes*, 113 (no. 633).

[86] Cf. Friedrich Schelling, Friedrich Hölderlin, and Georg W. F. Hegel. 2003. Oldest System Programme of German Idealism, trans. Andrew Bowie, contained in Andrew Bowie, *Aesthetics and Subjectivity*, Manchester: Manchester University Press, 2003, 334–335. The authorship of this unsigned fragment is contested, with many tending to think it is Schelling's work and others attributing it to Hölderlin. To think of it as a group expression of romantic views then held by these two and Hegel, all then friends, as Safranski does, is a workable solution – cf. Safranski, *Romanticism*, 48. This same view of the authorship is espoused by Adler and Louth, cf. J. Adler and C. Louth, 2009, Introduction, in Hölderlin, *Essays and Letters*, xi.

[87] Hölderlin later explored new mythology of reason themes in his poetic works and in his novella *Hyperion*; cf. J. C. F. Hölderlin, 2019, *Hyperion, or the Hermit in Greece*, Trans. Howard Gaskill, Cambridge: Open Book Publishers. Schelling discusses the new mythology in his *System of Transcendental Idealism*; cf. F. W. J. Schelling, 2001, *System of Transcendental*

mythology was to be pursued, under an "Idea which unites all", which is the "Idea of beauty[88]". Harnessed with the intuition, beauty could help-guide the thinking "I" toward understanding nature and the world of culture as one, and toward knowing them both (and knowing them both as one) via a combination of artistic intuition, rational philosophical speculation, and mathematics and science. Beauty would help "give wings again to our physics which is progressing slowly and laboriously via experiments[89]". The beautiful was also the means by which the new mythology could be evangelized; "The people" had to be converted to the new romantic way of seeing reality. *Systemprogramm* concludes:

> First of all I shall speak here of an Idea which, as far as I know, has never occurred to anyone – we must have a new mythology, but this mythology must be in the service of the Ideas, it must become a mythology of *reason*.
>
> Before we make the Ideas aesthetic i.e. mythological, they are of no interest to the *people* and on the other hand before mythology is reasonable the philosopher must be ashamed of it. Thus enlightened and unenlightened must finally shake hands, mythology must become philosophical and the people reasonable, and philosophy must become mythological in order to make the philosophers sensuous. Then eternal unity will reign among us. ... A higher spirit sent from heaven must found this new religion among us, it will be the last, greatest work of mankind.[90]

This new mythology of reason required the ability to intuit nature across all domains; as Ogden points out, Schlegel for example held that "without ... '*Vermögen der Naturanschauung*' [capacity for nature-intuition], the hope for a ... [new mythological view of nature] has no foundation[91]". Scientific knowledge would not be dry or reveal nature as a mere machine; it would be ennobling, artistic, and reveal nature as Nature's art-work. The new mythology of reason was also realist in the sense of seeking to intervene in the lives of "the people," to whom the

Idealism, Trans. Peter Heath, Charlottesville: University of Virginia Press. Schlegel discourses on it in his Talk on Mythology; cf. F. Schlegel, 1982, Talk on Mythology (within the Dialogue on Poetry), trans. Ernst Behler & Roman Struc, in *German Romantic Criticism*, ed. A. Leslie Willson, New York: Continuum. Novalis references it explicitly and implicitly in many of his fragments, novels and poems.

[88] Schelling et al., Oldest System Programme, 334.

[89] Schelling et al., Oldest System Programme, 334; also cf. the first section of this chapter.

[90] Schelling et al., Oldest System Programme, 335.

[91] Mark R. Ogden, 1989, Amor dei Intellectualis: Hölderlin, Spinoza and St. John, *Deutsche Vierteljahrsschrift für Literaturwissenschaft und Geistesgeschichte*, 63: 443.

new mythologists imagined evangelizing. We can also see in this early text the German romantic hallmark of emphasizing the affective quality of knowing; this, too, can be traced back to Spinoza, though Spinoza pursued in fact a different and more austere form of, in Pethick's phrase, "*making knowledge the most powerful affect*[92]". For Pethick, Spinoza's "knowledge *begins and ends in joy,*" "it is the task of philosophy to make knowledge the most powerful affect for living, transient existence," and the ideas of Spinoza only "make sense" once we understand the "*amor Dei intellectualis*" (or "intellectual love of God", cf. *Ethics*, V/Prop. XXXII, Cor.; Prop. XXXIII) and Spinoza's view that "the thought of *Deus sive Natura* [God or Nature] is the highest affirmation and offers the most intense affective experience possible[93]". Again, this speaks to Spinoza's own focus on, even contentment with, contemplating being as a monist whole. The romantics, for their part, also wanted to engage with the world as a place of growth and change and life, which they could then also place in dialectical tension with the eternal, infinite, absolute whole. A successful new mythology of reason must use beauty, and the capacity of art to furnish real [necessity-based] truth about reality, to translate the Spinozan *amor Dei* into terms which would make contemplation of God or Nature appeal to all, even those "unenlightened" not of the romantics' own educated, literary, creative class, so that all could experience the [highest] joy of knowing the world.

The new mythology was also part of the romantics' call for the unification of the sciences themselves,[94] as well as the unification of science with all other ways of speaking about nature, notably art and philosophy. In texts subsequent to the *Systemprogramm* the romantics continued their sacralization of the thinking subject seeking the unity of all via an elevated science and mathematics. Novalis writes that "Only after a passion for Nature has been bestowed on an entire nation, and a fresh bond fastened among its citizens – with natural scientific researchers and laboratories in every location – only then will we make any progress[95]". Schlegel urges the "study of physics, from whose dynamic paradoxes the most sacred

[92] Stuart Pethick, 2015, *Affectivity and Philosophy after Nietzsche: Making knowledge the most powerful affect*, Basingstoke and New York: Palgrave Macmillan, 3. Emphasis original.

[93] Pethick, *Affectivity*, 4–5, quotation from 5, 16.

[94] Robert M. Wallace, 1985, Introduction, in Hans Blumenberg, *Work on Myth*, trans. R. Wallace, Cambridge: MIT Press, vii.

[95] Novalis, *Notes*, 219 (nos. 15, 97).

revelations of nature are now bursting forth in all directions[96]". At the end of his famous "Talk on Mythology," Schlegel presents the concept of the new mythology thus:

> Be worthy of the greatness of the age and the fog will vanish from your eyes; and there will be light before you. All thinking is a divining, but man is only now beginning to realize his divining power. What immense expansion will this power experience, and especially now! It seems to me that he who could understand the age – that is, those great principles of general rejuvenation and of eternal revolution – would be able to succeed in grasping the poles of mankind, to recognize and to know the activity of the first men as well as the nature of the Golden Age that is to come. Then the empty chatter would stop and man would become conscious of what he is: He would understand the earth and the sun.
>
> This is what I mean by the new mythology.[97]

Nature, in this type of construct, is held open as a future knowledge-reward, *if* human beings can grasp their true destiny as knowers. The religious impulse is not to be left behind, but instead co-opted in the service of bringing the new mythology of reason into being. Schlegel exhorts, "let us reawaken all religions from their graves and through the omnipotence of art and science reanimate and reorganize those that are immortal[98]".

The new mythology of reason was basically an aesthetic project designed to elevate the romantic epistemological-ontological way of approaching nature and begin to, in the mode of "prophets" of the new mythology, lay out the new standards of meaning they wished to see established,[99] along with the correct unity-seeking methodology for following those standards and generating new forms of truth well into the distant future. It was not a revanchist program for recovering any particular myth-like story of nature for its own sake, and indeed its use of the category of myth also inevitably entailed a certain view of "old" mythology which was inherently overriding of it – including overriding of any ways the traditional creation myth in its proper context may actually exemplify a Jacobian circumscrib-

[96] Schlegel, Talk on Mythology, 101.

[97] Schlegel, Talk on Mythology, 101–102.

[98] Schlegel, *Fragments,* 96 (no. 22).

[99] Cf. e.g. Halmi on this point as it pertained to their use of symbols, Halmi, *Genealogy,* 18–19. For Halmi the romantics broke open the old idea of the symbol in the process of seeking a new "ideal of meaningfulness itself" and a more thorough ability to "[make] nature symbolic", cf. Halmi, *Genealogy,* 19.

ing of theoretical reason itself within the truly external world which utterly precedes and exceeds it.[100]

In the rubric of the new mythology of reason, "old" mythology became just a useful template for how creative poetry (which was how these romantics saw "old" myths, including Western monotheistic myths of creation such as the Genesis account) could both symbolize and directly create, or help uphold, a harmonious and unified society and knowledge-system. It was not anything to be merely read on its own merits or understood in its actual historical context or function to the best of our ability. Mythologies of past (and contemporary, foreign) societies offered a foretaste of the type of holism, the ambitious omni-valent way of knowing reality, the romantics wanted to remake in and for the post-Enlightenment West.[101] Schlegel writes, "All the classical poems of the ancients are coherent, inseparable; they form an organic whole, they constitute, properly viewed, only a single poem;" once the future age of knowledge arrives,

[100] There is no space here to link or compare a) Jacobi's conviction that existence and the real are wholly Other to theory, with b) some of the implicit, structurally-made claims of the traditional creation myth, but I will endeavor to explore this interesting question in future.

[101] It is not necessary to go too far into the complex and voluminous landscape of the study of myth (in the romantics and in general) to note simply that this potentially sets up tension with most other ways of reading "old" mythology. At a minimum, we can contrast the romantic gloss on old mythology with the readings of myth by many twentieth- and twenty-first-century scholars, such as K. Armstrong, or historian of religion Thrower, or philosopher of science turned historian of science Feyerabend. Thrower criticizes any failure to understand that myths actually represent "an entirely different apprehension of reality from that which now characterizes the predominant Western approach to the world" and rebuts any effort to read myth as "proto-philosophy or proto-science." Halmi, with Theisen, attributes just such a lens to these romantics: building on work on myth by earlier German scholars like Heyne, Schiller and Herder, including Heyne's idea of myths as "philosophemes," or "proto-philosophical causal explanations of natural phenomena, formed before language and reason had developed sufficiently to permit abstract thinking" (which had deeply influenced Herder as well), the romantics developed a view of myth, including the Biblical cosmogony, as a form of symbolic poetry, not allegory, as had been the predominant Enlightenment interpretation. Theisen argues that the romantics' approach to traditional myth patently constituted a "structural reconceptualization of antiquity under the aegis of modernity". Cf. Karen Armstrong, 2005, A Short History of Myth, Edinburgh: Canongate, 73–74; James Thrower, 1980, The Alternative Tradition: Religion and the rejection of religion in the ancient world, The Hague: Mouton, 17–18; Paul Feyerabend, 2016, Philosophy of Nature, ed. Helmut Heit & Eric Oberheim, trans. Dorothea Lotter & Andrew Cross, Cambridge: Polity Press, 89–90; Theisen, Romantic Myths, 9–10, 12; Halmi, Genealogy, 138–139; Hans Eichner, 1982, The Rise of Modern Science and the Genesis of Romanticism, *PMLA*, 97 (1), 19.

humanity will once again possess a *single* "eternally developing book", in which the "gospel of humanity and culture will be revealed[102]".

The idea of "old" mythology as exemplifying a harmonious capture and/or grounding of the key ligaments of an entire ancient culture (ancient Greece as well as non-Western cultures such as Egypt and India were the romantics' usual examples), and ultimately as deriving from (and reinforcing) some single unified worldview, informed their desire to use their new mythology of reason as a template for imagining a self-contained, naturalistic society of a future golden age, one in which all knowledge and art would "once again" form a single unified whole cultural reality. This would be a state of affairs highly preferable to the fragmented state of all the German arts and sciences around 1800, a state which the romantics felt must be changed for one which properly reflected and symbolized the overarching unity of all of nature (and of nature with culture).[103] Theisen points out that for these romantics, "old" myths showed "tautegorical" or "circular and self-referential" logic in action, including how it could ground a unified truth and art; they sought in their new mythology of reason to lay the groundwork for just this type of tautegorical logic to return to society in future – now based on reason and art alone and not mythical figures of gods and goddesses.[104] Again they try to take an objection of Jacobi's, that necessitarian, logical, causal reason is identitarian, and make a virtue out of what he considers an insurmountable flaw.

The new mythology of reason also sought to compensate for the loss – in the eyes of the romantics a loss which was regrettable and even tragic, but also undeniable – of a *Grundsatz* or first principle[105] for cosmological thinking, since the loss stung both logically and morally. A higher kind of physics must be found in part so it could reveal that a natural world intrinsically suited to moral action (and valuation) by human beings still existed. The human thinker must be shown how to find ultimate meaning in unity and necessity themselves, and our ability to think them; to find beauty and meaning through acknowledging one's *in medias res* situatedness, without firm foundations or anything toward which to orient one's life other than

[102] Schlegel, *Fragments*, 102–103 (no. 95).

[103] Halmi notes that one of the aims of early German romanticism was "counteracting intellectual disaggregation". Cf. Halmi, *Genealogy*, 146.

[104] Theisen, Romantic Myths, 9–10.

[105] Cf. Nassar, An 'Ethics', 238f.; cf. also Nassar, *Absolute*, 90f.

the way the earthly could appear under the auspices of the idea of the absolute unity of all.

The new mythology of reason exemplifies and extends the deeply transformative character of romanticism: its desire to transform the self, to reveal the physical world by transforming it into a single unified "nature" in which we fundamentally share, to transform all human history and culture into a single planetary evolution, to integrate all ways of knowing and forms of truth into a single higher unified view. Early German romanticism sought to transform existence, romanticizing the physical to the point of finally transcending it altogether by making it and/or knowing it entirely as knowledge and art, and these in turn one day, in the distant future, dissolving into a single, perfectly unified view. As noted above, the term "romantic" was deliberately explored by Novalis and Schlegel as bearing the sense of a transformative power, able to be applied to anything: an idea, an individual life, or an entire people, epoch, or natural expanse of time and space. For Novalis we must "raise" or "lower" the world in order to understand its meaning, writing, as already cited above, that "the world must be made Romantic" through "endowing the commonplace with a higher meaning, the ordinary with mysterious respect, the known with the dignity of the unknown, the finite with the appearance of the infinite[106]". The notion of a new mythology of reason was the expression of the romantic desire to establish the aesthetic grounding of this project of ceaseless transformation and knowledge-increase, so that it would self-perpetuate, and become something which lifted up, unified, and benefited human society. The romantics sought to present their Spinozist universe as inherently full of beauty and meaning, and to position themselves as harbingers of a second enlightenment to come. They tried to declare that modernity could be positively re-shaped; man could aspire to a substitute soteriological outcome via the new naturalism if it were properly animated by the intuition and the creative-artistic powers of the subject. This lastingly influential framework for the re-enchantment of the world through reason alone was deeply opposed to Jacobi's position. For Jacobi, in seeking to aestheticize the world made by reason, and further convince ourselves of its "reality" by trying to maximize its social importance and effects, we risked demanding the total and permanent relinquishment of all individuals' ability to "recoil" from it as a pseudo-world, integrating us into a larger society in the worst, most enchaining

[106] Novalis, 1997, Logological Fragments I, in *Philosophical Writings*, 60 (no. 66).

sense and threatening our last shred of awareness that the "world" fully explained and ontologized using theoretical reason is nothing but nothing.

As Josephson-Storm and Garrard, e.g., note, the romantic notion of a Europe spiritually and intellectually adrift because it lacked, or in the sense that it lacked, a real organizing mythology would itself become one of the most prominent legacies of the entire movement, later reshaped by Weber and others into the idea of a disenchanted late modern era that had begun with the Enlightenment, was first called out by the romantics, and entailed suffering from a lack which called out for filling[107], a lack which implicitly generated a permanent romantic call to arms to furnish a new Western mythos, albeit one built on reason and not religion as of old, a new mythos which would finally enable an intentional, conscious, and, again, reasoned re-enchantment of nature and world, now seen as ultimately the One or the infinite absolute. Such a new mythology of reason re-enchantment agenda, where the human would bestow qualities derived from its own imagination upon the world, re-making it, is the very opposite of Jacobi's call for a return to radical humility in the face of the actually external.

Locating a Tacit Jacobian Non-philosophy of Non-nature Opposing Early German Romantic Nature

The early German romantic philosophy of nature arose as something of an offshoot of the Controversy. Yet it is fairly straightforward to show how radically the philosophy departs and differs from what we might discern and draw out as a tacit Jacobian position on nature, or "non-nature", to negate the romantics' Spinozist use of the term, in keeping with Jacobi's other negations– non-knowledge, non-philosophy – meant to differentiate and demarcate the kinds of truths he sought from the knowledge prized by others, especially the new naturalists (cf. Chap. 1). As Hampton notes, there is a manner in which Jacobi's thinking as a whole bears relevance for

[107] Graeme Garrard, 2005, *Counter-Enlightenments: From the eighteenth century to the present,* London: Routledge Taylor & Francis, 61. Josephson-Storm holds that the romantics' version of the "clockwork" universe was a straw man they built only to tear down: "It was not [Enlightenment] physics that produced the nihilistic clockwork universe, but [early German romantic] philosophy," cf. Jason Josephson-Storm, 2018, *The myth of disenchantment: Magic, modernity, and the birth of the human sciences,* Chicago: University of Chicago Press, 75.

critiquing the "anthropocentric conceptualisation of reality[108]". We can also recall Livieri's gloss, per Chap. 1 above, that "the assumption that the whole of reality is but the activity of an immanent principle, as if thinking on the one side and nature on the other were just echoes of a formal self-generating whole," is a "key misconception" of naturalism[109] for Jacobi. As we have seen, for Jacobi both moral and logical problems must result, and in an intertwined way, from a romantic, or a follower of any other speculative philosophy of nature, really believing that some given knowable, principle-based, whole actually is the real world around us. These moral and logical problems are arguably also deeply environmental ones.

In the last, 1807–1812, portion of the Spinoza Controversy, the episode where he attacks Schelling, when Jacobi is older, and perhaps wearied from becoming, in his view, one of the only non-naturalists remaining in the "whole world" (see Chap. 1), Jacobi expresses his view of the superiority and crucial importance of the type of creative thought which must form one of the key capacities we preserve in and for [what I have been calling] his metarational subject. This creative thinking is that which is undertaken by the soul or spirit, i.e., those forms of thinking undertaken by the fuller person (not the theoretical faculty) which "do not have as their object what whiles away the time, but rather what brings it to a standstill and resolves it[110]". Such spiritual cognition is also always irreducibly individualistic[111] for Jacobi (as we can also recall from the previous chapter):

> Whenever we possess such cognitions, have them in control for actual use, we do it each time only to the extent that the spirit in each of us is capable of bringing it to life in us. Such cognitions cannot be distinguished from the actual power of the spirit, cannot be produced from outside nor confirmed from outside... It is in life that those cognitions must be grasped, that they must constantly be held. To be capable of this is the noblest and highest power of the soul—a power not to be found in any sort of Solomon's ring of ancient or modern philosophy that one need only purchase and put on, or in any talisman of some so-called religion that one puts on while perform-

[108] Alexander J. B. Hampton, 2023, Introduction, in *Jacobi and the Ends,* 2.
[109] Livieri, A Study, 7.
[110] Jacobi, On the Divine Things, 145.
[111] Cf. Jacobi, On the Divine Things, 145.

> ing prescribed rites. It is a power that the human being must invoke and awaken in himself.[112]

In its characteristics as time-stilling, and actually individually generated and experienced (as opposed to programmed or pre-set by others or by a received faith tradition), spiritual cognition can also establish a basis for a neo-Jacobian form of environmental [metarational] cognition. In the freedom of this form of thinking, which also must remain vital and life-based rather than rote, the subject gets actual powers restored to it: the power to choose which questions matter most to the self regarding the real world as such, and the power to be still, implying the power to be as an end in one's self and not only as one serving the aims or power (or systematized map of reality) belonging to others, and also implying a certain re-emergence to prominence of the space or place in which the subject experiences and relishes this stillness. These kinds of subjective powers are also highly conducive to enhancing, and focusing our attention on, the primal dialogue with reality which Jacobi consistently insists upon; indeed, as a living, breathing, life-embedded way of thinking, spiritual cognition must rely on our primal dialogue with the world-as-Thou; this dialogue is the oxygen to its fire. Thinking one's way to stillness is conducive to encountering the truly external as a Thou as this type of cognition has only spiritual curiosity as its directive and guide; it does not need to seek to raise the entropy of the world in somehow spending or consuming it, or using it as the means to an end. Jacobi further implicitly links theoretical reason (by contrast) with this more consuming approach to the world, and to our imposing ourselves upon it in some way, when he cites how our preoccupation with the natural laws which theory grasps does not reduce our "radical ignorance" but "rather ... *[distracts]* ... *[us] from it*[113]". Knowing via applying these laws, imposing our internalist naturalism as the real world, whiling away the time, and being distracted from our radical ignorance, are all part and parcel of the unreasoned nihilism which Jacobi associates with naturalism.

Jacobi's overall characterization of the act of rational world-building as inherently violent substitution (as discussed in Chap. 1) is another clearly environmentally meaningful position: the an-*nihil*-ation of the real is both figurative and potentially literal. This accusation of violence cannot be

[112] Jacobi, On the Divine Things, 145.

[113] Jacobi, On the Divine Things, 144, emphasis added.

assuaged by turning to the German romantics' quest for unity and oneness with the natural world, *pace* their poetic renderings (in response to both Kant and Jacobi) of the human subject's emotional and intellectual joy at experiencing glimpses of just such a direct unity. In Jacobi's framework, the romantic claim of the [cognitive] oneness of all reveals itself as the destruction of the sacred, grounding Other which is both really the precondition of thought, a bulk of raw being we will never fit into concepts, and also that which the human being needs to clearly signpost or mark out, rather than seek to subsume into the self. Without defining itself as different from this overweening reality, the theoretical faculty and even the metarational person will fall into a state which is irrational in a complete sense. In the Jacobian frame, romantic insistence upon oneness and unity between mind and nature becomes only a variation on Spinozistic naturalism's foundational nihilistic stance of refusal to acknowledge the real as it is outside of thought, and shows reason coming untethered from the real conditions under which it is employed (which even Spinoza was guilty of, per above). Oneness reveals itself as a *nihil* made by reason, an idol akin to the "Unitotality" discussed in Chap. 1. Pnevmonidou similarly notes that in Novalis' work, for example, the world-creation of the romantic "I" is implicitly violent, as it asserts the poet's own capacities as a kind of creative ultimacy that usurps the world and any true otherness in other human beings, including through using romantic art as the "method by which to gain complete access to the other[114]". For Rigby, the path of return to an imagined unity, just like a political utopian project which must destroy its rival contemporary political classes or systems in order to reach a desired end state, "enjoins a process of ceaseless, and potentially violent, transformation of the given," rather than protecting or recovering anything "within the givenness of the natural world[115]".

For Jacobi's position this violence is not just outwardly-directed toward the world and other people, but also a form of inner violence, since it upends what he holds must be the correct hierarchical relation between the fuller person and his or her theoretical faculty. We recall that Jacobi's "I declare that my reason, my whole inner self, flies into a rage before a representation of this sort, it shudders in horror and fright; I

[114] Elena Pnevmonidou, 2011, Veiled Narratives: Novalis' *Heinrich von Ofterdingen* as a Staging of Orientalist Discourse, *German Quarterly*, 84, 35 (for quotation); *passim*.

[115] Kate Rigby, 2001, Recovering from the Fall: The greening of modernity, *Journal of the Australasian Universities Language and Literature Association*, 96, 39.

recoil from it[116]" is his response to the weaving-of-weaving view of reality being substituted for the really existing external world; this response evokes internal suffering and pain on the part of his "reason, ... [his] whole inner self" – which feels anger and fear at the wound inflicted upon it as it is curtailed or overridden by theoretical reason alone, as well as pain in its conscience or moral sensibilty at witnessing a violent or wrong act being done to the world as it really is. Over a century later, in the age of the West's second Controversy iteration (the 1920s), twentieth-century philosopher Ortega Y Gassett claims that the tearing away of philosophy and reason from contemplating any but physics' type of problems left "the intellectual life of Europe" to "suffer" from "what one might call the 'terrorism of the laboratories'", a terrorism he, not unlike Jacobi, partly traces back to Kant, the one who "did not interest himself directly in the great cosmic problems, but with the imperative hand of a town policeman ... stopped all philosophic traffic – twenty-six centuries of metaphysical thought – by saying, 'Let all philosophizing remain suspended until this question is answered: How are synthetic judgments possible *a priori*?,'" referring really only to the synthetic judgments of physics.[117] These philosophers write of "recoiling" and "terrorism" because the malpractice of elevating theoretical reason too high within the person, over the rest of reason, comes at a real, even existential, cost to the soul, which must express this pain partly through emotion–partly, that is, using the language of experience, not the language of theory. Jacobi and Ortega Y Gassett are not interested in a mere difference between one philosophical school and another, but in the looming end of an entire faculty in the subject; what is at stake is an entire source of spiritual freedom, and, also, an entire set of potential experiences of the really external world, including experiencing the freedom to engage in asking what are, for the actual, living human subject, the truly compelling and ultimate questions. And such changes in, and losses of freedom for, the subject also, again, seem to imply a corresponding shift in how the subject is able to see and experience the non-human external world, as we will continue to see in what follows.

In a philosophical novel written in 1792, Jacobi further evokes the violence inherent in naturalism, violence with both an outer and an inner

[116] Jacobi, Letter to Fichte, 511 (no. 23).

[117] Jose Ortega y Gasset, 1964, *What is Philosophy?*, trans. Mildred Adams, New York: W. W. Norton, 45–46.

dimension. He describes how the subject who would rationally world-build, or completely rationally or theoretically ontologize what is, must first "cut himself off violently from nature and his own being, through the most wanton misuse *of the faculty of arbitrary designation*, this two-edged sword of truth and falsehood", and "must have seized his life by the root so to speak, in order to throw it from him[118]": this violence is the precondition for believing the world really is what and how he theorizes it to be. An environmental connotation clearly appears in the metaphor of the subject as originally being rooted, like a plant or tree, in the world, and world-replacement using theory involving a fatal choice to self-uproot and cut (or fell) the full living self.

Likewise, in *On the Divine Things* from the 1807–1812 portion of the Controversy, Jacobi does treat of animals and the natural world as full of blind instinct and the laws of force, but also writes:

> … the spirit that inside the human being rises above nature is by no means a spirit contrary or hostile to nature, it does not want to tear the man from the man: such a division would mean annihilation. Everything that exists, except God, belongs to nature, and cannot stand but in connection with it. … nature is the sum total of what is finite. Annihilating nature would entail the annihilation of creation. A foolish wish, but one that the sages of this world have expressed in the most different ways. … Nature is the beginning of all things.[119]

Though the positions of the "sages" come in many forms, to the extent they are annihilative of nature by trying to rise above it, they commit the same type of error, an error so grievous as to cancel their right to be considered sages at all.

We can also discern, perhaps, implicit Jacobian affirmation of the extraordinary value of the non-human world where he calls the supernatural something which "cannot be understood through concepts" and "cannot be apprehended by us in any way except as it is given to us, namely, *as fact* – It IS!"; though he here refers to God,[120] this recalls and parallels his position that the most speculative philosophy can say about really existing

[118] Jacobi, Edward Allwill's, 495 (no. 316).
[119] Jacobi, On the Divine Things, 183.
[120] Jacobi, Concerning the Doctrine, 376 (nos. 426–427).

nature, or reality as a whole, is "whatever is, is[121]". Both forms of the acceptance of existing reality as brute fact, as raw is-ness which cannot be theorized, follow from the same basic imperative, which is to recognize the significant inherent limit (or even intrinsic defect) of our causal rationality's way of ontologizing using necessity-relations. Existence as such should always be perceived as making theory go into a state of mute pointing. Nature, as the other form of being aside from God, for Jacobi, would seem to potentially partake of the same rightful safety from conceptual designation and control, the same requirement that the fuller, metarational person approach it and encounter it primally (not conceptually), and recognize and recall it as truly a Thou.

Even where Jacobi notes that nature lacks the good and the wise, the majestic, and is the site where only the "law of strength" and "violence" prevail, nature is also an "abyss", with its hint of infinity;[122] it is indeed a "negative infinity", with no limits on its ability to begin and end, and of "unfathomable character" which refuses any rational or philosophical proof of its raw origins (or its meaning, outside, for Jacobi, a theistic frame); nature simply *is* one of only two types of being,[123] which per above are God and nature. Whether one uses a so-called naturalist's or a so-called theist's arguments, for Jacobi, as long as you seek to argue the question using logical relations of cause and necessity, products of our mere theoretical understanding, it is "equally impossible [in either theistic or naturalistic terms] to explain the existence of the universe from a First conceived as an origin[124]"; the natural world—like God, for Jacobi – is simply Other to the form of reason that seeks to use logical necessity to ontologize – to list, index, and explain – what is real.

Jacobi's response to the "weaving of weaving" view, which entails recoiling in horror, is also relevant for delineating a prospective [neo-]Jacobian environmental cognition departure-point in another sense. Environmental cognition as a form of Jacobian metarational or spiritual cognition would require us to move through the external world opposing its removal and substitution. It would not be a question of establishing a rational set of criteria for how to treat nature, in which situations; it would

[121] Jacobi, Concerning the Doctrine, 372f. (nos. 416–422), especially Jacobi's footnote to no. 419 on 373–374.

[122] Jacobi, On the Divine Things, 185.

[123] Jacobi, On the Divine Things, 185.

[124] Jacobi, On the Divine Things, 188.

be more akin to an intuited and preconceptual way of being. As we recall from the foregoing, spiritual cognition does not allow the subject to choose or weigh from a disembodied, calculative perspective, or receive cues from anything other than the individual soul in conversation with the really existent world in which it finds itself; environmental cognition would encounter the environment as something which stands outside and prior to our thinking of it, and thus as a worthy target of the free soul's consideration in itself, in addition to being worthy of recognition and high valuing as the intertwined oxygen forming part of free spiritual cognition's fire, as noted above. Jacobi expressly mentions the animals in his recoiling in horror representation of how he responds to the new world "made" of forces (and expressed through necessity-relations) alone, writing, as we recall, "what would a mere weaving of weaving amount to, from the animals below up to the saints above?—I declare that my reason, my whole inner self, flies into a rage before a representation of this sort, it shudders in horror and fright; I recoil from it.[125]" He shudders to see "even" the animals replaced in their reality by this "weaving of weaving" view. Jacobi's position counters the claims made by the German romantics and other Spinozists that deep pleasure, moral value, and even a kind of personal honor or bravery inhere in the experience of thinking nature through a monist, reasoned, self-contained lens. He is not content to cede the guidance of the affective sense (or the affective state) to the Spinozist view; Spinoza's famous *amor Dei intellectualis* (*Ethics*, Part V, Prop. XXXII/Cor.; Prop. XXXIII), to the extent it implicitly raises up Spinoza's system (and all logical and theoretical knowledge possessed by human beings) as a starting point toward experiencing such love, would be, for Jacobi, a kind of spiritual sickness or pitch of self-delusion, where we seek to love an image of our own making instead of the actual God. So, too, would an "intellectual love" of nature be an error of the soul.

Jacobi also arguably implicitly frees the external world from the physical interventions we base on our perfected, certainty-oriented necessity relations when he speaks (as noted above) of how Kant (and others) helped him understand that far from being a magical and *real* key to the world, the principle of necessity is not a final principle nor even a real "principle" at all; it only reflects the choice to establish some assumption as a premise and perform certain types of operations upon it. Jacobi's stance, *contra* the German romantics, that necessity relations accessible to and mappable

[125] Jacobi, Letter to Fichte, 511 (no. 23).

by theory cannot have final ontological meaning, has direct relevance for how the subject can attempt to most profoundly know the non-human world, including what type of knowledge of the non-human world counts as ultimate. In addition to bringing immediate consequences for the subject's intellectual worldview, this also by extension shapes what the human considers permissible to do *to* nature and world. If we come to see that our internalist "world" of theory is arbitrarily constructed, not ultimate or even really real *qua* self-upholding or complete world, it ceases to conceal from us that we, as humans, are always only "Thous" for other living things, acting in what really are inscrutable and arbitrary ways for them unless (and until) we seek to communicate with them or at least encounter them in the metarational, Thou- to- Thou sense. If as such a Thou for other life forms, our own species subjects them to annihilation and replacement, to pain and suffering and even extinction, the moral weight of such acts cannot be, as it were, sloughed off through the sense that we act in an authorized, masterful manner, uncovering and following the higher and more objective ways (or laws) of the single real world which we know and the other species cannot; the moral weight of our acts instead adheres to us; it becomes inescapable. We become unable to un-see that we are experienced by other species through the ways our world-building arbitrarily–for our world-building is exposed as immediately arbitrary for them just as it is also exposed, from the metarational perspective, as arbitrary for us–annihilates them; we must face that we are not experienced or encountered through our own inner worlds of theory, and cannot any longer assume the license to assert our theoretical world as the one real world via intervening force and technological conversion. This ability to see ourselves from the Thou-centered viewpoint we could, as Jacobian environmental cognizers, be said to share with all life requires that the metarational faculty be permitted to survive. In this sense, the metarational is also our most animal capacity, even as it is also, for Jacobi, our faculty most conducive to communion with the divine.

We can recall from Chap. 1 how Jacobi expressed shock that the full human person could ever be satisfied with the "mere schema" that remains after the human mind has destroyed the world as it really is, in order to re-create it as he can fully comprehend it. As the environmental corollary of this, we could point to the difference the subject experiences when it moves from a maximally degraded and synthetic environment – usually the result of our imposing our theoretical world in place of the real one – to a relatively intact and flourishing natural zone which has not been similarly used [at least not in the immediate past]. The subject apprehends the difference as an experience, and not via analysis or concepts. The natural zone is relatively full of life and meaning; it is a place we can stop to truly see it as it is, and be happy not imposing our concepts upon it. As for the zone of the domesticated, the synthetic, or the informatic, where we have maximally imposed and constructed our "worlds" made sheerly of necessity relations, we arguably find only, in Jacobi's words (as noted above in Chap. 1), "*purely self-intending and self-contemplating activity ..., without purpose, or point of arrival*[126]", requiring we effectively remove our full subjecthood and render ourselves mere generic understanders of which necessity-webs were mapped and used to build the zone, simply in order to navigate and survive in it.

As we have been discussing, belief in our own ontologies of necessity is not a trivial or incidental kind of belief for Jacobi; he implies that if one tries to really inhabit the pseudo-worlds of reason for long enough, the subject is put at risk, and may forget there is any other kind of activity in the real world other than the purely self-intending, self-weaving kind we "find" (or superimpose through force) there. The fuller subject risks altogether forgetting the real as all opportunity for spiritual or environmental cognition is removed. This in turn has multiple onward spatial correlates, from the desire to avoid non-anthropogenic spaces or even the outdoors altogether, to hyper-consumption enabled by both the subject's growing belief that the schema-world is all there is, and thus there is no moral obstacle to following its dictates exclusively and making our one existence on earth merely part of its mesh, and by the subject's constant need to "fill" its schema-world with manufactured objects since it is actually so empty of life and meaning.

A potential Jacobian "non-philosophy" of "non-nature" could be seen as re-grounding and re-acknowledging the need for environmental cognition, re-opening the subject's ways of relating to the real biosphere by removing the nihilistic practice of committing to see it as sheer necessity. Instead of the fuller self and the entire biosphere being placed in captivity by the theoretical capacity's necessity relations-based woven "worlds," a Jacobian non-philosophy of non-nature could position the fuller person as understanding that its full subjectivity begins before concepts, opening up new possibilities of freedom originating in its primal dialogue with

[126] Jacobi, Letter to Fichte, 511 (nos. 22–23).

non-nature as Thou, allowing the external non-human world to regain its Other and Thou statuses for us, and us to regain these statuses for it.

The radical ignorance we have, for Jacobi, is another potential impetus toward, and plank within, environmental cognition. Perhaps in the context of his insistence on this ignorance, his declaration that "scientific proofs" or any other truth-claims built on the ontologizing of necessity-relations "as" the real world can only ever "elucidate cognition, but not increase it in content or ever secure a ground for it"[127] (see next paragraph), may make Jacobi sound anti-science. But he was simply resisting science's self-interpretation as providing access to what "made" (and continues to make) the really existing world as such be present and exist; he was not opposed to science as long as it was interpreted within a metarational, realist frame. As he notes in the 1807–1812 instalment of the Controversy, he wants to oppose the de facto Kantian elevation of scientific reason over all other forms of reason. Kant himself clearly acknowledged the limited quality of mathematics and logic, their core identitarian character; for Jacobi (as noted earlier) it is incoherent for Kant to not then do more to actively preserve the rest of our higher, fuller, metarational reason and grant it the hegemony over theory which it deserves (cf. Chap. 1). He wishes Kant had "kept steadfast in mind what scientific proofs can and cannot do[128]". For Jacobi, Kant's claims to limit the access to reality as such enjoyed by theoretical reason (since theoretical reason cannot touch or encompass things as they are in themselves[129]), correctly call out its definitional quality, and so forth, seem inconsistent with Kant still effectively destroying extra-theoretical ways of being and knowing which even Kant put outside the ken of the theoretical understanding (cf. Chap. 1). For Jacobi, Kant's path leaves the subject marooned apart from the real (where the subject actually is in fact), and is thus still a nihilistic one. He also implies violence to the real world itself is thereby also done, at least cognitively, since it is left apart from us and fading away into immateriality [for us]. As noted in Chap. 1, in Jacobi's view, with Kant:

> ... nature, which up to that time was conceived as objective, vanished for good with its essence and all its works; once it was separated from external

[127] Cf. Jacobi, On the Divine Things, 169–170.

[128] Jacobi, On the Divine Things, 169.

[129] Cf. Kant, *Critique of Pure Reason*, 1781, e.g., Book I/ Second Part/ First Div./Bk II/ Ch. III.

> sense perception, it became nothing for the philosophizing understanding. In the face of the faculty of [logical] cognition, everything, the knower and the known, dissolved into an empty activity which imagined imaginary things: objectively, into a pure nothing.[130]

For Jacobi, the either/or is always inescapable: "necessity"-tracing causal reason nihilism, replacing what is with an identity statements-based weaving of weaving of our own making, or Jacobian externalism, which strives not to replace but to encounter and reveal. The first scenario generates a subject who may come to lack any reasoning capacities beyond repeating the axiomatic relations it has been conditioned to consider the real world, while the subject who takes the second path retains full personhood, one in which a causal reasoning capacity is still embedded, but is kept – as a result of reason's ability to really see and understand the inherent limits of this form of thinking – contained, used in specific times and for specific tasks, but never as the basis for ontological claims about what being (or all that is) actually is. The first choice for Jacobi cannot but lead to a "nothing everywhere manifest," or the "manifest nothing", immersing the subject in a cycle of self-reinforcing nihilism as this "nothing" replaces "God and all that is suprasensible or supernatural[131]", or, we might again add, truly external, prior, Thou, or Other, including the actual, infinite, universe and the earthy biosphere within it, as they really exist, and thus as they are super-natural (or super-schema) in the sense of being over and above the naturalist's necessity-defined, rationalized, constructed version of the natural.

Jacobi wrote in *Letter to Fichte* that the negative space left over outside the reasoned, constructed and only imagined world put together by theoretical reason as it world-builds [using relations of necessity] – since the world which reason can in fact succeed in constructing is so paltry, its efforts leave mostly void – could be an "infinite nothing, a pure-and-total-in-and-for-itself" that is both a nothing (from Jacobi's position), and yet something about which the "I" seems to know a great deal and to which it appears to attribute a number of naturalistic properties, since it imagines it as a real infinity but also as non-threatening to the I's world as known and knowable (as arguably is the case with the romantics' absolute). He then reasons that he must be allowed to counter a philosophy entailing

[130] Jacobi, On the Divine Things, 172.
[131] Jacobi, On the Divine Things, 174.

such "knowledge of the nothing" with his "philosophy of non-knowledge,[132]" a philosophy meant to [one day] provide truths about which types of knowledge are not possible – which types of knowledge are in fact non-knowledge – or, put another way, carve out a new kind of knowledge for philosophy to pursue that was anti-theoretical regarding nature as a whole, and thus needed to be defined through negating this growing trend in Western thought. Since in the aforementioned void, which theory leaves untouched and yet claims to encompass in its knowledge-schema nonetheless, Jacobi finds "nothing confronting me … except nothingness, and even chimeras are a good match for that," then he would not be "vexed" if Fichte, "or anyone else, were to call Chimerism the view I oppose to the Idealism that I chide for Nihilism.[133]" Jacobi's chimera speaks to human beings' fallibility and alloyed character. But it can also speak of a return of the human being to its full, animal, metarational and really living self; of a return of freedom, freedom to call out the *nihils* of science for what they actually are, instead of shrinking our conception of nature, mind, and reason to try to fit within their narrow compass; freedom to oppose the intellectual and physical domination of the real living world by their arbitrary designations.

Without stressing the metaphor too far, the chimera is also suggestive of the generative power of real life and real inventiveness, which derives from nature and not our minds and might therefore appear fearful, paradoxical, or nonsensical by logic's standards. The *nihil* fails to capture anything true about being though it trucks in vaunted conceptual assertions about the infinite; it is opposed by the ungainly, creaturely chimera, full of inexplicable vital force.

Jacobi writes in a portion of the *Letter to Fichte*:

> God is, and is outside me, a living, self-subsisting being, or I am God. There is no third. Were I not to find God outside me, before me, and above me, so that I have to posit Him … then I am myself this so called being.[134]

For Jacobi, God is/I am God is another either/or we cannot avoid; this is a theme which will return in Barth in the next Controversy iteration. There is potentially a tacit environmental extension of this same

[132] Jacobi, Letter to Fichte, 519 (no. 39).
[133] Jacobi, Letter to Fichte, 519 (no. 39).
[134] Jacobi, Letter to Fichte, 524 (no. 49).

dichotomy we can unearth, given all we have been discussing of Jacobi's willingness to characterize [non-]nature as real and non-conceptualizable being. If we were to experiment with paralleling the above quotation in environmental terms, we might obtain the either/or that either the external, real, non-human world is, or I am the external, real, non-human world, something akin to:

> The non-human world is, and is outside me, a realm of living, self-subsisting being, and beings, or I am the non-human world, I am the infinite universe and all its real beings. There is no third. Were I not to find the infinite world outside me, before me, and above me, so that I have to posit the world [and so place it, considered as being and beings as such, as entirely Other to analyzable relations of necessity] … then I am myself this so-called infinite world.

Theism, we might begin to discern, is not the only type of Jacobian spiritual, or metarational, cognition; it does not exhaust what can flow from a capacity to encounter the external as such.

From the foregoing, we can see Jacobian environmental cognition begin to emerge as a counter to early German romanticism's efforts to elevate the image of the natural world that might be thought to maximally result from their new Spinozistic philosophy of nature. Still, it remains for us to continue to unearth the Jacobian position on [non-]nature. While Halmi claims early German romanticism left a kind of dissatisfaction or even failure in its wake; that this philosophical effort burned out (by around 1808), in effect, after having been unable "to secure any actual meaning from a purportedly infinite store of potential meaning" in the world as they conceived it,[135] its basic philosophy of nature would dominantly live on, especially in the natural sciences. Since these sciences would then increasingly become the West's sole source of its understanding of nature or reality, the early German romantics were in a sense the most significant "winners" of the first Spinoza Controversy, just as their intellectual heirs would also carry the day in the Controversy's twentieth century reprise, to which we will turn next.

[135] Halmi, *Genealogy*, 18–20.

CHAPTER 3

A Second Spinoza Controversy? Neo-romantic Monism versus Crisis Theology in Interwar Europe

A century after Jacobi's death in 1819, the war-shocked Weimar Republic and neighboring countries reeled with what Weller calls cultural vertigo, a sense that something vital to society had been lost,[1] potentially permanently. As part of a response to this sense of crisis, German-speaking (and other) Europeans explored "a range of resacralizing modernist projects,[2]" many of which essentially involved new versions of German romantic re-enchantment aims, and many of which intertwined with a new round of enthusiasm for Spinozan and Spinozistic ideas which also returned in force to German-speaking Europe in this era. At the same time, the Christian "crisis theology" of the Swiss theologian Karl Barth emerged on the scene with a radically different response to the time, parts of which reprised some of Jacobi's positions in a new way.[3] The contrast between crisis theology and the resurgence of Spinozism produced what historian B. Lazier calls a return of the Spinoza Controversy's "configuration" to the world of German letters.[4]

[1] Weller, *Modernism*, 17f.

[2] Weller, *Modernism*, 4.

[3] Lazier, *God Interrupted.*

[4] For Lazier it is the "Pantheism Controversy" which has returned; he notes that for him the Pantheism Controversy refers to the 'series' of debates that stretched the last two decades of the 1700s, cf. Lazier, *God Interrupted,* 69. See Chap. 1 of this work regarding the the-

K. C. Snow, *Our Inevitable Third Spinoza Controversy*,
https://doi.org/10.1007/978-3-032-03045-0_3

Lazier's interest in this opposition turns on reading it as an appearance of two rival "heresies" both offering radical alternatives to a stunned post-war German-speaking world. He uses the 1932 congress in The Hague celebrating (the Dutchman) Spinoza's three hundredth birthday as a kind of focal point of the new Controversy's Spinozan (or Spinozistic) side,[5] with Carl Gebhardt, the "foremost Spinozist of his day," organizing the gathering. The crisis theologians were not in attendance, but by this point their opposition to the Spinozists had been clear for over a decade, ever since the publication of Barth's influential work *Epistle to the Romans* in 1919, a year before Gephardt and his allies founded their new Spinoza Society. For Lazier, from the outset of the "philosophical and theological configuration of the 1920s", Spinozism and Barthian theology (which Lazier likens to gnosticism) were offering "competing solutions to the same set of problems[6]". The new Spinozans (or Spinozists), who included those embracing Spinoza philosophically and those embracing him theologically (including both Christian and Jewish scholars), were largely following the version of Spinoza who was the "philosopher 'drunk with God,' invented by the German romantics a century before and copied by their Jewish followers,[7]" in this way continuing the liberal theological tradition started by the "pantheistic" German romantic theologian Schleiermacher and his successors.[8,9] Barth and his colleagues, on the other hand, "made clear their opposition to any notion of a readily intelligible divinity[10]", and rejected any notion that the divinity the romantic Spinozistic view had left as its legacy, now being picked up again, was or could be real.

matic similarity among all three episodes of the Controversy, and my use of Spinoza Controversy as the umbrella term for all three.

[5] Lazier, *God Interrupted*, 67ff.

[6] Lazier, *God Interrupted*, 70; 68.

[7] Lazier, *God Interrupted*, 85.

[8] Schleiermacher was in the wider German romantic circle and was one of those with whom the early German romantics discussed in the first two chapters regularly engaged; Ogden notes for example the shared interest by Schlegel, Hölderlin, Novalis, and Schleiermacher in the writings of St. John, and in creating a type of "Johannine Spinozism," cf. Ogden, Amor dei, e.g. 459.

[9] Lazier, *God Interrupted*, 68. For a taste of Schleiermacher's approach to theology and his weaving in of neo-Spinozistic, romantic themes, cf. e.g. Friedrich Schleiermacher, 1996, *On Religion: Speeches to its cultured despisers*, trans. and ed. Richard Crouter, Cambridge: Cambridge University Press.

[10] Lazier, *God Interrupted*, 68.

While I am indebted to Lazier's reading for drawing my attention to this interwar recurrence of the Spinoza Controversy "configuration", in his helpful term, and concur that the crisis theologians take up some of Jacobi's positions albeit in a different, more theological than philosophical key, I would also offer that we can broaden our consideration of the Spinozism prevalent in the era beyond those who consciously sought to revive his school of thought philosophically or embrace him within the Schleiermachian liberal theological tradition. Other, adjacent, new forms of what were essentially early German romantic positions – positions on and approaches to nature, in particular – were fluorescing in the Weimar cultural and intellectual milieu and next door in post-Habsburg Vienna. Taking Jacobi's understanding of the externality of the real to theory *contra* the Spinozistic early German romantic philosophy of nature as our slightly broader (and more environmentally-keyed) Controversy template per Chaps. 1 and 2, Barth and his colleagues can be considered in opposition to a wider swathe of neo-Spinozist naturalism re-emerging all around them (and overlapping with the theological and philosophical resurgence of the romantic version of Spinoza which Lazier details), including in areas where German romantic thinking was returning as part of ostensibly very secular and newly pared-down forms of philosophical and scientific modernism (e.g., in the Vienna Circle, per below). To provide further historical context on this return of such a [broader] Spinoza Controversy "configuration" (to stay with Lazier's term), it is worth offering here a brief sketch of the history of some of the main tributaries feeding into this eventual resurgence in the 1920s of neo-romantic ideas of nature, beginning where the original Spinoza Controversy leaves off.

The Intertwining of Early German Romantic "Nature" and Western Science 1812–1939

Several historians have taken note of various aspects of the lasting influence of early German romantic philosophy of nature upon European and North American science. Caneva holds it is difficult to prove, and Gower that it is evidently untrue, that Schelling and his romantic contemporaries directly influenced the development of modern physics as a whole.[11]

[11] Gower says Schelling's was an "abortive attempt to revolutionize scientific thought" which petered out in ideas about galvanism, and in Oersted's theories of dynamical chemistry: cf. Barry Gower, 1973, Speculation in physics: the history and practice of *Naturphilosophie, Studies in History and Philosophy of Science,* 3: 327.

Caneva highlights Hans Oersted and the minerals scientist Christian Weiss as two scientists directly influenced by Schelling who also made important contributions to their fields.[12] Some of the papers of James Clerk Maxwell suggest an early German romantic view of or approach to nature.[13] Abrams points to a sizeable influence of *Naturphilosophie* on nineteenth and twentieth century physics (mostly referring to electromagnetism and field theory).[14] Wilson claims the Schellingian, and broader romantic, approach to nature had "epoch-making consequences for the subsequent development of scientific cosmology and of physics in general" and cites romanticism's influence on Oersted and Faraday as examples.[15] Gale claims that modern biology as a whole "emerged from *Naturphilosophie's* inklings of cell theory[16]". Richards calls "the historical or developmental approach" of the early German romantics a "leitmotiv that will resonate through much of subsequent Romantic literature and science", and notes Schelling was admired by Alexander von Humboldt and many other natural scientists of the nineteenth century, also making a case for the influence of early German romantic and Goethe-ian approaches on Darwin.[17]

Separately from any question of direct German romantic lineage, the development of the theories of evolution and geology themselves during the nineteenth century offered a number of new and tempting openings for future Spinozists to attempt to combine their concept of nature with the empirical claims of these new fields. Darwin's theory of natural

[12] Kenneth L. Caneva, 1997, Physics and '*Naturphilosophie*': A Reconnaissance, *History of Science*, 35: 97, 54.

[13] Cosmologist Wilczek cites an extract from Maxwell's papers wherein Maxwell writes that man must partake of infinity and strive to be an "impersonation of the divine processes of nature, and to show forth the union of the infinite with the finite"; he elsewhere cites Maxwell on how the individual wanting to cultivate an awareness of justice and truth in his own existence can draw on the laws of nature for support, cf. Frank Wilczek, 2015, *A Beautiful Question: Finding nature's deep design,* New York: Penguin, 164, 192.

[14] M. H. Abrams, 1973, *Natural Supernaturalism: Tradition and revolution in romantic literature,* New York: W. W. Norton, 171.

[15] Wilson characterizes Oersted's belief in a unifying force in nature as romantic and notes that this belief led to his 1820 work on formulating electromagnetism. He also casts Faraday's perception that not only electricity and magnetism, but also gravity, could be joined as one, as romantic, also noting it was a view which prefigured Einstein's. Cf. Andrew D. Wilson, 1993, Romantic cosmology, in *Encyclopedia of Cosmology,* ed. Norriss S. Hetherington, New York: Garland, 602–604.

[16] George Gale, 1975, Leibniz, Chew, and Wheeler on the Identity of Physical and Philosophical Inquiry, *Review of Metaphysics*, 29: 323–324.

[17] Richards, *Romantic Conception,* 109, 129.

selection through time seemed particularly congenial to those favoring a Schellingian paradigm of a time-bound, self-grounded, intelligible and developmental Nature-nature with its own inner principles and powers giving rise to a gradual complexification of life out of simpler forms. Darwin's theory also seemed, to the romantically inclined, to bring subject and object together in new ways; the human was shown to be part of the same evolutionary web as all other forms of life, evoking a powerful new sense of unity between mind and nature. Surveying such vast expanses of time and space as was required to conceptualize the geological formation of the Earth itself and the evolution of all plants and animals as well as man, and the new vistas of the night sky being opened up by Herschel's and others' newly powerful telescopes, all while removing or at least challenging the role of divine telos and other aspects of God's role in nature (as these had been understood up through the Enlightenment period), also made the threat of horizontalized, totalized materialism remain the dogging shadow of the more optimistic and future-directed romantic spirit as it tied itself in with these new sciences.[18]

Early German romantic thought had an influence, too, both upon subsequent iterations of German romanticism from later in the nineteenth century all the way into the twentieth century,[19] and upon other romantic movements, including those in Britain and the United States. These romanticisms in turn then also directly engaged with, influencing and being influenced by in turn, German and Anglophone science and philosophy. Early German romantic thought, particularly via Schelling, directly influenced the early English romantics Wordsworth and Coleridge.[20] As historians of environmentalism Hinchman and Hinchman note, for Coleridge, like Schelling, "no part of nature is fully intelligible until one has understood the whole[21]". Raiger notes that Coleridge held that "the true poet and true scientist know that they see the world through symbols", and this "yields a recognition of the unitary form of human

[18] Cf. footnote 45 in Chap. 2.

[19] Safranski, *Romanticism*.

[20] Coleridge and Wordsworth had spent a year in Germany 1799–1800 and were exposed to German romantic works including Schelling's, cf. e.g. Garrard, *Counter-Enlightenments*, 63.

[21] Lewis P. Hinchman and Sandra K. Hinchman, 2007, What we owe the Romantics, *Environmental Values*, 16: 337.

subjectivity and laws of nature, subject and object[22]". Gare notes that Coleridge himself had a close relationship with many scientists and mathematicians, and contends that "to a considerable extent, the most creative developments in science" since Coleridge's era "can be seen as progress of the Romantic conception of nature as consisting of activity producing and maintaining stable forms, fields of force and self-organizing processes in place of Newtonian cosmology[23]". Wordsworth expressed his sensitivity to mathematical beauty;[24] the journal *Nature*, launched in 1869, which would become the preeminent scientific journal in the world, for nearly its first century in print carried a masthead containing a quotation from Wordsworth implying an eternal and monistic whole "Nature" of the romantic type was the nature which the journal's scientific findings sought to directly reveal.[25]

Along with the term "nature's" new Spinozistic connotations, the romantics also bequeathed to subsequent ages a newly naturalistic version of the old Greek term "cosmos", which entered the lexicon primarily via the German romantic author Alexander von Humboldt, who published his multi-volume work of popular universal natural history, *Cosmos*, in the middle of the nineteenth century at the end of Humboldt's long life and career as a scientific explorer, amateur geologist, ecologist, and botanist: according to Sachs, it was Humboldt's book that was at least partly

[22] Michael Raiger, 2010, Coleridge's theory of symbol and the distinction between reason and understanding: A genealogical recovery of the Baconian method of science, *History of European Ideas*, 36: 315.

[23] Arran Gare, 2014, Deep ecology, the radical enlightenment, and ecological civilization, *The Trumpeter*, 30: 192. N.B. however that Newton did not interpret natural laws in the same way as the early German romantics.

[24] Writing, e.g., of how, through a study of geometry, he might "beguile his sorrow, and almost/Forget his feeling", and not only "So was it with me then" but also "so will be/With Poets ever", cf. Carl Stahmer, 2005, *Romanticism, hypertextuality, and metavisual information theory*, ProQuest Dissertations & Theses Global, 68, citing Wordsworth *Prelude* 6.135–187.

[25] The masthead quotation read: "'To the solid ground/Of Nature trusts the mind which builds for aye.' – Wordsworth". The line ran on *Nature's* front cover (originally under a pictograph of the word "Nature" floating in the sky above the Earth amid a comet and planets) from its inaugural issue in 1869 until the late 1950s, and stayed on the title page after that until 1963. "For aye" here means "for ever" – or "for forever" in contemporary usage. The line changes and implicitly secularizes Wordsworth's original text subtly; his original line had been "To the solid ground/Of nature trusts the Mind that builds for aye." Cf. Thomas Owens, 2015, *Nature*'s Motto: Wordsworth and the Macmillans, *Notes and Queries*, 62: 430–435.

responsible for bringing the term "cosmos" back from the Greek into regular German (*Kosmos* was its German title) and English usage[26], imbuing it with new romantic connotations in the process. Both indirectly through Wordsworth and Coleridge, and via the continued spread of ideas directly from Germany, early German romanticism also influenced the views of nature held by later English romantics such as Keats, Shelley, and Byron. According to Stahmer, Wordsworth and Shelley both conveyed their interests in linguistic theories, and in unities between mind and nature (and unities in nature), to Ada Lovelace, who may have been influenced by their thinking as she carried out her own work related to early concepts of computer programming.[27]

In the United States, early German romantic philosophical conceptions of nature continued to grow and evolve into new forms with the burgeoning of American romanticism and transcendentalism. Ralph Waldo Emerson, who also read Schelling, sought to, in the words of one historian, "'grasp the infinitely large' by piercing the natural form and reducing it to a set of principles," and perceived a single governing intelligence at work in nature which meant "the mind could unlock the universe", at least in principle, though Emerson's thinking (like all romantics) also

[26] Aaron Jacob Sachs, 2004, *The Humboldt Current: Avant-garde exploration and environmental thought in nineteenth-century America,* ProQuest Dissertations Publishing, 9, note 20; 133. On p. 133 he is partly citing Robert K. Barnhart, *1995, The Barnhart Concise Dictionary of Etymology: The Origins of American English Words*, New York: HarperCollins, 165.

Kosmos was not only, or not necessarily, a physical space or place as much as it referred to the *fact of ordering,* a *having-made-orderly* which seemed to have somehow transpired. Originally, *kosmos* had a "double meaning 'order' and 'world-order'"; cf. Daniel W. Graham, 2010, trans. & ed., *The Texts of Early Greek Philosophy: The complete fragments and selected testimonies of the major Presocratics, Part I,* Cambridge: Cambridge University Press, 239.

Kosmos in the Greek sense was delineated from what was called *apeiron*, or not-formed, indefinite, or without-boundaries, in the Presocratic cosmologies. *Kosmos*, then, in the Greek sense was located within (or contrasted with) *apeiron*, in an undefined way. When Humboldt brings the term "cosmos" back into use, interestingly, the *apeiron* is gone, and the cosmos conveys a sense of order alone, orphaned from any context.

[27] Cf. Stahmer, *Hypertextuality.* Related interests were held by Novalis, who desired for example to join new German teaching on mines and minerals with grammar, including via a study of individual words, cf. e.g. Novalis, *Notes,* 49 (no. 333); or Novalis, *Notes,* 195 (nos. 9, 11), where he writes, "The numerical system is the *model* for a genuine system of linguistic signs – The letters of our alphabet shall become numbers, our language, arithmetic," and "The applied mathematical sciences show us Nature as a mathematician."

exhibited some tension on this point.[28] Day notes that Emerson thought it necessary to utilize a direct intuition of nature "in isolation, in the wilderness[29]" in order to know it; as Lanza partly points out, in isolation the poet could then uncover that he and nature were co-grounded in God, in something close to the German romantic, pantheistic or panentheistic, sense.[30] Per Neubauer, Walt Whitman, like Coleridge, Wordsworth, and even Shelley to some extent, saw a unity between art and science.[31] American romanticism and transcendentalism also exercised their own influence on the Anglophone (largely American) pragmatic school of philosophy, which in turn affected American science and scientific epistemology. (Pragmatism was also influenced by romanticism via logical empiricism, cf. below.)

Anglophone romanticism also preserved and augmented early German romantic approaches to universal natural history like Humboldt's (and even Schelling's own). As Zakariya notes, by the mid-1800s, some scientists, mostly within biology and geology, were increasingly espousing the possibility of a "universal history" in ever more ambitious terms. Zakariya points to John Tyndall's 1874 address to the British Association for the Advancement of Science (BAAS) in Belfast as displaying a conception of universal natural history rooted in German and Anglophone romanticism; he also notes that Tyndall (and the scientist Helmholtz) embraced a kind of "transcendental materialism" as a philosophy of nature.[32] Romantics on both sides of the Atlantic had read Humboldt's broadly published *Cosmos*,[33] first translated into English in 1849, which was a natural history

[28] Jennifer Baker, 2007, Natural Science and the Romanticisms, *ESQ: A Journal of the American Renaissance*, 53: 407, also referencing Laura Dassow Walls, 1995, *Seeing New Worlds: Henry David Thoreau and Nineteenth-Century Natural Science*, Madison: University of Wisconsin Press, 113, 1.

[29] Aidan Day, 2012, *Romanticism: The new critical idiom*, London: Routledge, 192; also citing Deborah Madsen, 1998, *American Exceptionalism*, Edinburgh: Edinburgh University Press, 74.

[30] Robert Lanza, 2007, A new theory of the universe, *The American Scholar*, Essays: 33; he is referring to Emerson's *The Over-Soul* (1841).

[31] Cf. John Neubauer, 1993, Science and Poetry, in *The New Princeton Encyclopedia of Poetry and Poetics*, eds. A. Preminger et al., Princeton: Princeton University Press, NPN; he is referencing the Preface to Whitman's *Leaves of Grass*.

[32] Nasser Zakariya, 2017, *A Final Story: Science, myth & beginnings*, Chicago: University of Chicago Press, 127. Per Zakariya, Tyndall had been influenced by the "romantically minded Thomas Carlyle" as well as Faraday; cf. Zakariya, *Final Story*, 98.

[33] Alexander von Humboldt, 1858, *Cosmos: A Sketch of a Physical Description of the Universe*, trans. E. C. Otté, London: Henry G. Bohn.

that included, or at least gestured toward, the history of the universe as a whole as well as the entire planet Earth, now portrayed as a dynamic, changing, interconnected unity that was both connected outward to the universe (and its history), and contained and shaped all human societies and their histories, implicitly giving them a shared common grounding. Edgar Allan Poe attempted his own idealistic and intuitive answering "history" of the universe, *Eureka* (1848), which he dedicated to Humboldt.[34] *Eureka* was ostensibly a cosmogony and cosmology in "prose-poem" form, but it is somewhat unclear whether *Eureka* is celebrating or lampooning Humboldt, since *Eureka* reads something like a combination of a romantic universal natural history combining idealism with realism, a fanciful poem, and a withering diatribe about overreach by both philosophy and science. It was widely panned by critics at the time, perhaps because they were unsure whether Poe meant the work to be taken seriously or ironically;[35] in the more ironic (and more likely) reading, a Poe-ian version emerges of Jacobian skepticism toward universal-scale ontologizing using relations of necessity and analyzable causes, and Jacobian resistance to using the narrative device of an objective historical account to bolster this ontology as correct and ultimate.

In the American pragmatic school of philosophy, C. S. Peirce developed a kind of complex cosmology influenced by Schelling (and Hegel), an "objective idealism[36]" which included an idealist cosmogony from nothing. His cosmological thought presaged some aspects of later big bang scientific cosmologists' (neo-romantic, as we will discuss) interest in explaining the laws of nature themselves as the result of pure chance operating within some type of overarching framework of "necessity". Peirce was very engaged with the question of how any primordial homogenous chaos could possibly transform itself into a law-driven and diversified cosmos (one of the many paradoxes which arise when monist naturalism is

[34] Edgar Allan Poe, 1848, *Eureka: a Prose Poem,* New York: G.P. Putnam.

[35] Mihai A. Stroe, 2014, Poe's hieroglyphic universe: The master keys, *Romanian Journal of Artistic Creativity,* 2: 19–81. Stroe notes that Poe was, incidentally, one of the favorite authors of Alexander Friedmann, the Russian cosmologist who was the first to solve Einstein's cosmological field equations; it is not known if Friedmann read *Eureka*.

[36] The pairing "objective idealism" is highly evocative of the German romantic quadratic nature discussed in Chap. 2.

taken to be literally true of the universe, and one which continues to confront big bang cosmology today, as we will see).[37]

Reaching the European interwar period and the resumption of the Spinoza Controversy "configuration," in Lazier's term (cf. above), two main channels within these broader ongoing continuations of early German romantic currents can be identified as most pertinent: the overlapping schools or approaches in philosophy variously known as the Vienna Circle, logical empiricism, and logical positivism, on the one hand; and, on the other hand, and closely related to these philosophical developments, the new physics, with its theories of relativity and quantum mechanics, and after 1917, its new scientific cosmology.

The Monists in Interwar Europe

The overlapping philosophical movements or schools referred to as the Vienna Circle, the logical empiricists, and the logical positivists, were influential, mostly interwar, European philosophical currents tied in closely with German and [transatlantic] Anglophone scientific developments (and with some strands of American pragmatist thought, as noted above). For Bowie, early German romantic philosophy played a "subterranean role" in the group's thinking overall;[38] I would argue that the philosophy of nature which emerges in this grouping is actually fairly patently neo-German romantic and neo-Spinozistic.

As Ryckman notes, the Vienna Circle (and associated schools) rose to prominence in the early twentieth century by claiming the authority to translate, as it were, the twin behemoths of the new modern physics, relativity and quantum mechanics, to the rest of the academic (and broader) world.[39] They largely did so using a framework of a new type of monism and a combination of idealism with realism. Philosophers in this group like

[37] Cf. Claudien Tiercelin, 1998, Peirce's objective idealism: a defense, *Transactions of the Charles S. Peirce Society*, 34: 1–28; and T. L. Short, 2011, Did Peirce have a cosmology?, *Transactions of the Charles S. Peirce Society*, 46: 521–543. The twenty-first century big bang cosmologist Lee Smolin cites Peirce's understanding of natural laws (and 'habits' of natural things) approvingly, cf. Lee Smolin, 2015, Temporal Naturalism, *Studies in History and Philosophy of Modern Physics*, 52: 96–97.

[38] Andrew Bowie, 2000, The Romantic Connection: Neurath, the Frankfurt School, and Heidegger, Part I, *British Journal for the History of Philosophy*, 8: 276.

[39] Cf. Thomas Ryckman, 2005, *The Reign of Relativity: Philosophy in physics 1915–1925*, Oxford: Oxford University Press.

Moritz Schlick claimed that the substantive novelty, and empirical success, of general relativity (GR) constituted incontrovertible evidence *contra* Kant's claims that seemingly foundational concepts in science like space and time were merely our own mental categories (not externally discovered) and that reason, including scientific reason, was unable to access nature as it is in itself.[40]

In many ways the Vienna Circle and associated currents were using and recreating a version of the early German romantic quadratic "nature" (cf. Chap. 2) which equated epistemological access to nature via science and mathematics with direct, and even in-principle final, ontological insight or capture, and did not refrain from asserting the nature they were able to thus access was both ideal and real (this was also particularly true in the realm of Einsteinian cosmology, as we will see below). Though their style was more propositional and austere, they were neo-romantic in their assertion that they could "think nature" directly and *in medias res* in a language increasingly shared between physics and philosophy, and in the sense too that their monism excluded any other possible speech of nature as legitimate. They evoked Novalis and Schlegel in their views of mathematics and logic as real-ideal languages whose truths were of the highest or the most ultimate type available to human subjects. And both mathematics and physics constructed reality via expressing relations of logical necessity now once again perceived as holding (self-identically – deploying a neo-romantic understanding of symbol[41]) ultimate existential, ontological referential capacity and meaning. Arguably, given the sparseness of their theoretical depiction of nature, these thinkers were also always tacitly relying on a version of a Spinozistic, romantic Nature simply being posited as always-present (and always with productive power within it, and always bound by necessity alone). This allowed them to touch or invoke real being, simply by uncovering or creating propositions which could be shown to be true by the standards of mathematical and logical necessity. Historically, once these approaches to philosophy were eventually (after

[40] Cf. Ryckman, *Reign,* 6, where he is also partly quoting Schlick speaking in 1922: Since the subject "'finds itself obliged to use non-Euclidean geometry" to describe the "same world [as Newton's]", "Through Einstein ... [it] has now become a reality, [that] the Kantian position is untenable, and empiricist philosophy has gained one of its most brilliant triumphs'". Kant holds time and space to be "the two primitive *quanta* of all our intuitions," not things we encounter in nature directly, cf., e.g., Kant, *Critique,* Book I/ Second Part/ Second Div./ Bk II, /Ch. II., /Sec. 1.

[41] Cf., e.g., Halmi, *Genealogy*, p. 63.

World War II) criticized for, among other things, hiding their tacit framework, philosophy of science did appear in many ways to simply unearth the Spinozistic Nature that had been there all along, and carry on with the same basic assumption that it was valid (and/or unavoidable) to use some construct like this, never really challenging the fundamental logical empiricist view that reality could be successfully captured via this hybrid way of merging mathematical physics with an assumed infinite natural well of being manifesting only necessity and never requiring any articulation, description, or explanation other than what mathematical physics provided and philosophy helped elucidate. The logical empiricists were self-avowedly (mostly) monists (a new journal called The Monist had launched in 1888, in which some of the Vienna Circle published), and their view of nature had an inbuilt ease with meshing with a new version of a single universal Nature-nature of a Spinozist type, again, whether this was tacit or, later, after the Second World War, gradually more openly embraced, particularly as cosmology advanced as a science.

While the Vienna Circle and logical empiricism schools, and those who overlapped with them, can be divided into multiple disparate groups in any traditional history of philosophy, in the context of considering the return of the Controversy during this era, they coalesce as roughly one anti-Jacobian and neo-romantic position on nature. While some philosophers in these groups did refrain from claiming the Kantian era over, more either showed Schlick's triumphalism or sought to circumvent the issue of highest truth or ultimate reality (and our access to it) altogether by embracing the status of the human subject as inevitable world-creator in realist-idealist mode, and the conflation, once again, of epistemological success with ontological finality.

Even if "all" the subject could speak about was precise observations via science, and this positivism was held to be non-cosmological in its aims, in its exclusion of all other speech of nature, logical positivism, like logical empiricism, was still de facto totalizing and monist. The preoccupation with representing reality via logical and mathematical ways of clearly and precisely showing relations of necessity, increasingly in an intertwined way with mathematical physics, also led to an increased focus by these interwar philosophers on analyzing human language and the structure of truth-claims in both science (and mathematics) and philosophy itself. But even with the attention to language, the same slippage as the romantics had embraced between "it cannot be otherwise" referring only to our systems of logic, and "it cannot be, otherwise" invoking existential and ontological significance (cf. Chap. 2), was not policed, and was even embraced[42]; or to the extent it was ostensibly policed by some, tended to only generate a claim that philosophy must dismiss any and all speech of being or existing *tout court*, since this topic was too metaphysical for the new philosophy, which must be one with science. In a way this merely silenced the question, without removing the need for even the most positivistic among these thinkers to tacitly rely on a simply posited or assumed omnipresent Spinozistic Nature-nature which their crystalline scientific-philosophical propositions could simply reach out and touch, or directly, romantically symbolize, justifying the elevated interpretation given to such statements that they were the only type needed in order for human knowledge of reality to advance and deepen. Thus the end result across the spectrum of interwar logical empiricism was the erection of a monist system with no seams. And in the work of the more triumphalist among them, the romantic idea that scientific study supported by mathematical logic really could tap into the one real unity, nature itself, or that nature itself simply was relations of necessity, quite clearly made a return. The heavy focus on epistemology as ontology helped enhance these philosophers' interpretation and elevation of the experience and activity of knowing the world of nature via propositional and logical truths based on experimental and other sense data, and justify philosophy's redirection to addressing the task of considering the new scientific truths and how best to interpret them. As is suggested in the very term logical empiricism, often what was interpreted as most significant was our "empirical" interactions with truths built of logical necessity-relations, both in themselves (again, giving rise to a growing focus on analyzing human language as well as mathematics internally) and as science allowed us to link them with nature (and even effectively constitute nature solely out of them).

In the interwar era many European thinkers and artists sought glimpses of some better future for Western subjects which did not resemble the horrors of the continental war they had just endured. These philosophers were no exception, putting forward an aesthetic self-interpretive project which joined with the "resacralizing" aims of the traumatized society, per Weller (cf. above), and which in many ways recalled and re-presenced the German romantics' idea of advancing a new mythology of reason.

[42] Per Scruton, the Vienna Circle and schools connected with it tended to use G. Frege's way of indicating existence via logic and logical symbolization; cf. Roger Scruton, 1995, *A Short History of Modern Philosophy: from Descartes to Wittgenstein*, London: Routledge, p. 241.

The Vienna Circle held that science could, as Ryckman notes, become "the primary instrument of human advance from the dreary annals of superstition, dogma, and fanaticism that permeate human history[43]". As Romizi has also documented, the "scientific world-conception" consciously espoused and cultivated by the group gave rise to an active public and political ideology, particularly after 1929 when group members published a kind of manifesto[44] which distilled one thing that united the otherwise non-homogenous group: simply the stance that there *was* such a thing as the scientific world-conception, that it was necessary to actively promote it as such, and that it constituted some kind of ethical or normative good for society at large to do so.[45]

Among the romantic new mythology of reason-like goals which flowed from this position was the goal of the unification of the sciences themselves, as well as the bringing of other fields like philosophy fully under science's umbrella. There was an aspect of this which reflected Novalis's dream of a single, unified, real-ideal understanding of reality which incorporated all human knowledge. They worked on expanding the use of mathematics and logic in science, and rendered philosophy, Creath points out, part of the "scientific enterprise[46]". The scientific enterprise continued to gain in cultural power, including as a result of these unifying steps; the interwar period arguably saw the first peak of what Ortega y Gasset characterizes (as noted in Chap. 2) as the Kant-initiated shift toward taking human scientific and logical thought itself as the most important target of philosophical inquiry, moving away from entertaining any questions lying fully outside science, and moving toward instead placing philosophy

[43] Ryckman, *Reign*, 12.

[44] The pamphlet was published in 1929 as *Wissenschaftliche Weltauffassung: Der Wiener Kreis* (*The Scientific Conception of the World: The Vienna Circle*), Wien, Artur Wolf. It had no stated author, but the Preface was signed "For the Ernst Mach Society" by Hans Hahn, Otto Neurath, and Rudolf Carnap. The text is reprinted as O. Neurath, 1973, Wissenschaftliche Weltauffassung: Der Wiener Kreis, in *Empiricism and Sociology*, eds. M. Neurath, and R. S. Cohen, Vienna Circle Collection Vol. 1, Dordrecht: Springer. https://doi.org/10.1007/978-94-010-2525-6_9

[45] Donata Romizi, 2012, The Vienna Circle's 'Scientific World-Conception': Philosophy of science in the political arena, *HOPOS: The Journal of the International Society for the History of Philosophy of Science*, 2, 212f. Also cf. Michael Stöltzner, 2015, The Logical Empiricists, in *The Oxford Handbook of Causation*, eds. Helen Beebee, et al., Oxford: Oxford University Press, 108.

[46] Cf. Richard Creath, 2017, Logical Empiricism, in *The Stanford Encyclopedia of Philosophy*, ed. Edward N. Zalta, NPN.

wholly in the service of exploring and helping to address scientifically-framed problems—or even questions about science's own internal consistency and proper interpretation, with the rise, also out of logical empiricism (and barely distinguishable from it), of the new sub-branch of philosophy known as the philosophy of science.[47]

The new mythology-like social and political goals expressed in the 1929 manifesto were indications that the philosophico-scientific enterprise self-interpreted its new power as highly warranted, and hoped for more to come in the future. It was unapologetic about its authors' goal of ending metaphysics: the "Scientific World-Conception" pamphlet stated that

> there is no realm of ideas that stands over or beyond experience. Nevertheless the work of 'philosophic' or 'foundational' investigations remains important in accord with the scientific world-conception. For the logical clarification of scientific concepts, statements and methods liberates one from inhibiting prejudices ....[48]

What they mean by "experience", as the realm they want to re-assert as primary, is here really scientific *experiment*. The authors claimed to want to "confidently approach the task of removing the metaphysical and theological debris of millennia," which they noted might also be thought of as them "returning, after a metaphysical interlude, to a unified picture of this world which had, in a sense, been at the basis of magical beliefs, free from theology, in the earliest times.[49]" This is highly reminiscent of the romantics' own interest in magic (cf. Novalis' mention of the need for "magical realism," per above), and their dream of a return to a unified worldview

[47] Cf. above and also e.g., Ortega y Gasset, What is Philosophy, 78–79. Western philosophy's interest in "analytic" philosophy more broadly began well before World War I, and involved continual exchange and co-influence between especially Germanophone and Anglophone philosophers as a sizable portion of them turned their attention toward trying to better understand the structure and function of language, mathematics, science, and logic, and away from metaphysics or what Ortega y Gasset might consider traditional philosophical topics and questions. Ortega y Gasset's "terrorism of the laboratories" characterization (cf. Chap. 1) can be understood to attach to this shift in its entirety. The Germanophone inter-war monist philosophers I am discussing here were one example of, and a very influential part of, this larger shift.

[48] Wissenschaftliche Weltauffassung, 316–317.

[49] Wissenschaftliche Weltauffassung, 316–317. The interest in magic as the forerunner to science is also reminiscent of Novalis, who spoke of his view of nature as entailing "magical realism" or "magical idealism," cf. Wanning, The Poet, and also cf. Chap. 2.

like (but also unlike, because now wholly reason-based) the ones ostensibly enjoyed by past societies as shown by and in their "old" mythologies. The logical empiricist worldview aimed, again, to reject any notion that there were questions which philosophy as science-independent metaphysics could still address which science (or the new philosophy of science) could not; as Creath notes, any impulse to reserve a domain in which philosophy alone could consider "matters, perhaps cultural ones ... more profound and important" than those the scientific world conception could legislate, and any belief that "fidelity to evidence and punctilious argument is somehow small-minded", both had to be intentionally banished.[50]

In the way these philosophers spoke of nature, as in the early German romantic approach, there were no first principles, and no "being" to worry over naming,[51] since the "world" accessible to and constructed by science was always-present, and propositions could simply be assumed, a priori, to make direct [neo-romantic] symbolic reference to it. In a way which went well beyond what the German romantics would have attempted or sought, these philosophers effectively removed all that was external and real (Jacobi might say) except for what was knowable via scientific experiment and mathematical logic, in order to world-build using epistemology-as-ontology and placing necessity relations squarely in the center of what counted as nature. Cosmography or the "scene" of the physical world per se was not their primary target but exposing those laws and behaviors that were now cast even more ambitiously as nature's innermost structure, and this to some extent was also true of their indirect colleague Einstein even as he developed new cosmological applications for general relativity.

Einstein's friend, the mathematician Gödel, was in the Vienna Circle, and the Circle's (and related groups of) thinkers were also, per above, directly engaging with Einstein's and other physicists' new theories of nature, theories about its behavior which covered everything from the very smallest scales to the very largest. Einstein, like the Circle's philosophers, regularly cited the philosopher and scientist Ernst Mach (b. 1838–d. 1916) as a major influence. Mach worked on causation and its representation in mathematics, echoing and enhancing an idea also, per above, explored by the early German romantics, that nature's behavior was intrinsically mathematical, allowing mathematical depictions of causes in nature to be somehow directly, romantically, symbolic of nature itself in

[50] Cf. Creath, Logical Empiricism, NPN.
[51] Again cf. Scruton, *A Short History*, 241.

a realist way – a position which would reverberate throughout science and philosophy ever after – and also offering notions, as Pojman points out, which were partly rationalistic and partly based upon evolutionary theory, of the need to conceive of nature as a single causally connected whole[52], notions which Einstein would later seek to develop and test mathematically, including in his new interpretation of the laws of gravity as being able to offer the world its first truly scientific cosmology.[53]

The logical empiricists' neo-romantic position that the new scientific worldview should allow human beings to reliably create not only a knowable world but one with values and inherent value, a morally sufficient and desirable world, would remain deeply ingrained in physics ever after. Ryckman assesses that it was the claim by the Vienna Circle that relativity and quantum mechanics had re-established science's empirical access to nature in itself, contra Kantian restrictions (a claim that other philosophers of science would later question, but to little effect on the sciences' own self-interpretation), that formed the main foundation of later twentieth-century scientific naturalism, with all its connotations of offering something like a complete worldview.[54] Życiński notes that the "cognitive monism" of the Vienna Circle was formative for modern physics, linking it with thinkers like Stephen Hawking: he calls this monism the belief that there is "one rational interpretation of the world that eliminates any sense of mystery", that "science can answer all ultimate questions of humankind", and that physics is able to "provide explanations that ... [are] at the same time consistent and complete[55]".

The romanticization of intellectual work as endless progress is typical of the Spinozist side in all three Controversies. In this interwar period, infinite horizontal collection of data and insights leading to "a genuine axiomatization of science" was one of Vienna Circle philosopher O. Neurath's

[52] Cf. Paul Pojman, 2023, Ernst Mach, in *The Stanford Encyclopedia of Philosophy*, ed. Edward N. Zalta & Uri Nodelman.

[53] Cf. Christopher Smeenk, 2014, Einstein's role in the creation of relativistic cosmology, in *The Cambridge Companion to Einstein*, eds. M. Janssen and C. Lehner, Cambridge: Cambridge University Press, 228–269; and Pierre Kerszberg, 1989, *The Invented Universe: The Einstein- de Sitter Controversy (1916–17) and the Rise of Relativistic Cosmology*, Oxford: Clarendon Press.

[54] Ryckman, *Reign*.

[55] Józef M. Życiński, 2002, Metaphysical and epistemological presuppositions in Stephen Hawking's interpretation of the creation of the universe, *Roczniki Filozoficzne* (*Annals of Philosophy*), 50: 126, 125.

imagined goals, per Ziche;[56] this again evokes Novalis and his colleagues. The notion of intellectual work, especially science and its interpretation, as the foundation of a future age of enlightenment, and maximally ennobling of human life, also characterized the interwar philosophers' neo-romantic new mythology of reason-like claims.

Einstein's New Scientific Cosmology

A little over a century elapsed between the end of the original Spinoza Controversy and Einstein's 1917 paper, "Cosmological Considerations in the General Theory of Relativity," which launched modern scientific cosmology as a discipline and served as another clear sign that neo-German romantic, neo-Spinozistic ideas of "nature" had returned to the West in earnest, if they had ever left.[57] Einstein was famously a self-described believer in Spinoza's God,[58] though how exactly he interpreted this remains slightly unclear, and drew inspiration from Spinozism in his cosmologizing ambitions, and in his view that science was the only religious activity that remained possible in his era.[59] Einstein called himself an

[56] Paul Ziche, 2012, Monist philosophy of science: Between worldview and scientific meta-reflection, in *Monism: Science, philosophy, religion, and the history of a worldview*, ed. Todd Weir, New York: Palgrave Macmillan, 164.

[57] Albert Einstein, 1952, Cosmological Considerations in the General Theory of Relativity, trans. W. Perrett and G. B. Jeffery, reprinted from *The Principle of Relativity*, eds. H. A. Lorentz, et al., Dover, 175–188. Available at http://einsteinpapers.press.princeton.edu, Vol. 6: The Berlin Years: Writings, 1914–1917 (English translation supplement), Doc. 43.

[58] Cf., e.g. Enrico Giannetto, 2009, The Electromagnetic Conception of Nature at the Root of the Special and General Relativity Theories and its Revolutionary Meaning, *Sci & Educ*, 18: 765–781. On 777 Giannetto discusses how Einstein is known as a follower of Spinoza, one for whom doing science was "contemplation and knowledge of Nature-God". There is also the well-known story of the 1929 telegram to a rabbi in New York who had asked him if he believed in God; Einstein replied: "I believe in Spinoza's God, Who reveals Himself in the lawful harmony of the world", cf. Gerald Holton, 2003, Einstein's Third Paradise, *Daedalus*, 132: 31.

[59] E.g., Gleiser cites approvingly how "Einstein wrote that the devotion to science is the only truly religious activity in modern times"; Einstein himself, in a speech honoring Planck, said of Planck's "state of mind" which allowed him to do "work of this kind" that it was "akin to that of the religious worshiper or the lover." Cf. Marcelo Gleiser, 2005, *The Dancing Universe: From creation myths to the big bang*, Lebanon: Dartmouth College Press, xii; and Albert Einstein, 1982, Principles of research, Reprinted from Albert Einstein, *Ideas and Opinions*, trans. Sonja Bargmann, New York: Crown, 227. Available at http://einsteinpapers.press.princeton.edu, Vol 7: The Berlin Years: Writings, 1918–1921 (English translation supplement), Doc. 7.

"unscrupulous opportunist" in matters of epistemology rather than a "systematic epistemologist", again reminiscent of German romanticism's emphasis on trusting to our intuition and our innate ability to think nature itself, as well as its conviction of the underlying harmony among all forms of true knowledge; Einstein combined elements of realism, idealism, positivism, Platonism, Kantianism, and, again, Spinozism, in his development and interpretation of his new, relativistic physics,[60] and then cosmology.

Einstein's 1917 cosmological paper was dedicated to testing general relativity (GR) at limit conditions of [mathematical] infinity, in order to challenge the too-"mechanistic" Newtonian universe, with its rather undefined cosmology,[61] and replace it with a relativistic one conceived via a unified, solely scientific, model of the universe as a single, Machian-Spinozistic, mathematically representable entity.

His conceptual shift from the infinite or indefinite mathematical grid space (or absolute space and time) type of universe of Newton (and Descartes) to a new conception of the universe as a finite yet smooth and unbounded continuum[62,63] is what is usually remarked upon as marking a paradigm shift in the history of Western cosmology; but far more significant in our Controversy context is simply his move from Newton's reticence to use empirical science and mathematical demonstration to attempt some form of raw ontology without theology,[64] to Einstein's own asser-

[60] Thibault Damour, 2005, Einstein 1905–1955: His approach to physics, in *Einstein, 1905–2005: Poincaré Seminar 2005,* eds. T. Damour, et al., Basel: Birkhäuser, 152. Einstein is here quoted from another source, Paul Arthur Schilpp, 1949, *Albert Einstein: philosopher-scientist,* Lassalle: Open Court, NPN; Damour notes he is drawing from remarks called "'Reply to criticisms' which he [Einstein] wrote for the [Schilpp] book," cf. Damour, His approach, 151.

[61] Historian of science Kerszberg characterizes Einstein's two main motivating objections to Newtonian cosmology (or quasi-cosmology) as, first, Einstein's conviction that "there is no unique centre" to the universe, and, second, that "the infinite is not a 'place'", cf. Kerszberg, *Invented,* 363; cf. also Smeenk, Einstein's role, 239ff., on this point; Smeenk notes that for Einstein, Mach's principle helped him believe that spatial conditions at infinity, or infinite boundary conditions, should not be able to "do" anything, i.e., influence matter, cause inertia, etc.

[62] Cf. Einstein, Cosmological Considerations, 427.

[63] There is an interesting immediate echo, in Einstein's "finite yet unbounded" universe concept, of the German romantic conception of the finite world as the limit of the infinite ceaselessly making it; cf. Chap. 2.

[64] Cf. e.g., Newton's Query 31, in the *Optics,* where he makes clear he works from phenomena and leaves to divine causation the origins of the first particles, cf. Isaac Newton, 2014, Queries to the *Opticks,* in Isaac Newton, *Newton: Philosophical Writings,* ed. A. Janiak, Cambridge: Cambridge University Press, 184f.; or Newton's General Scholium to the

tion that this could in fact be done (though as we will see, he still arguably tacitly relied on an offstage Nature eternally creating an always-present nature).

As one well-known astronomer in the interwar years would note in critiquing this move, Einstein's field equations did not really change the equipment available to the scientist from what had been available to the scientist in Newton's era in any categorical way; they were still just rules for making measurements of gravity, not any more implicative of any one universe-object or -place than Newtonian mechanics.[65] As philosopher of science Ryckman similarly notes, the theory of GR consists only of the recipe for constructing the "simplest conceivable relation of comparison in a[ny] continuum[66]". What changed in 1917 was, then, rather the interpretation the scientist began to place on statements of natural laws or equations governing, in this case, the pressure of a perfect spherical solid (which was roughly how Einstein modeled the universe in this paper, albeit in four dimensions – we will return to the model below), taken, again, to an absolutely generic limit condition represented by mathematical infinity. The new view was that natural laws (or other scientific statements about consistent necessity relations observable in the behavior of matter and energy) could, particularly when geometricized, somehow symbolize something constitutive of the entire world in which they held true, even though this same world was also seen as having somehow "made" these laws be present, or instantiated them as true and bearing, in the first place.[67] Natural law X (and any of its applications in various imagined conditions) could be used as a shorthand for "world in which natural law X holds," particularly since the laws now ostensibly pertained in part to how "spacetime" itself could behave, spacetime being now a real feature of reality or nature again, as Einstein and others rehabilitated space and

Principia, where he discusses the universe's intertwining with God, and also takes pains to distinguish his view from Spinozist-type (not properly Spinozan) pantheism where there is no final causation and no separation of God from our rational causal explanations, cf. Isaac Newton, 1999, *The Principia: The Authoritative Translation: Mathematical Principles of Natural Philosophy*, trans. Bernard Cohen, Anne Whitman, and Julia Budenz, Berkeley: University of California Press, 586ff.

[65] Herbert Dingle, 1937, Modern Aristotelianism, *Nature*, 139: 785.

[66] Ryckman, *Reign*, 242–243.

[67] Again, for Newton, arguably, while the laws of gravity and motion did symbolize something quite foundational about space and time themselves, he did not present these laws as having *made* space and time, nor as able to symbolize the universe as a whole.

time back from Kant's diminishment of them to mere inbuilt cognitive frames for human knowledge (cf. above).

The instantiation of a raw beginning of our gravitationally-bound universe was initially avoided in the 1917 paper, which modeled the universe roughly as an eternal and static four-dimensional spheroid. Soon, however, Hubble's astronomical observations interpreted as cosmic expansion were conjoined with Einstein's exercise (cf. below), and an origin point of a certain kind for our observable universe was added to the model. Now the universe we saw around us began in the ultimate "simplicity" of the big bang singularity and gradually grew more complex, in a way highly evocative of Schelling's idealized natural history, in which the cosmic past is simple and elementary, and then gradually through time nature grows more complex, to culminate in the present and in human minds. This romantic view of natural history, which had deeply influenced new sciences like geology and evolutionary biology (and been later influenced by them in turn), had relied, in its original German romantic articulation, on a then-new kind of inverted Neoplatonism, in which emanation (out of an always-present Nature) was developmental, and types or entities of things in nature built up toward increasing complexity (and by implication finally toward reunion with the One) along a horizontal or temporal axis of becoming. This general view was not far from how Hölderlin and others saw historical dynamic laws of nature applying to both human history and to the evolution of the physical universe. This new[68] type of Neoplatonism dovetailed with the romantics' aversion to a creation or origin-point in nature in the traditional sense. Eichner calls this novel type of romantic progressive Neoplatonism a "monumental" shift away from longstanding Western presumptions that "higher" had always created "lower," usually beginning with God; for Eichner, Schelling upended the order of things so that "the 'higher' developed *from* the 'lower,'" also necessitating that "the world was not created once and for all by a perfect, supreme being but has grown or developed[69]". As Beiser points out, in the context of universal natural history, this was presented as the emergence over time of

[68] Kenney notes however that some aspects of the romantics' Neoplatonism also partly reflects Plotinus' own teachings as well, cf. John P. Kenney, 2005, *The Mysticism of Saint Augustine: Rereading the* Confessions, New York: Routledge, 37. Goethe also seems to support this idea of a 'lower' strata of parts shared by human consciousness and the natural: cf. Richards, *The Romantic Conception*, 439.

[69] Hans Eichner, 1982, The Rise of Modern Science and the Genesis of Romanticism, *PMLA*, 97: 15.

various levels of complexification out of the same "living force[70]". Schelling saw as a "fundamental task of all nature philosophy: TO DERIVE THE DYNAMIC GRADUATED SEQUENCE OF STAGES IN NATURE" (his capitalization), so as to contribute to the representation of nature as a single "universal organism".[71] This romantic approach[72] left two legacies that are here picked up by scientific Einsteinian cosmology in the new wave of romantic constructions of nature, now fueled by the new physics, which emerge in the interwar years: the connective link between infinite Nature and finite historical nature, where Nature seems to cause nature, but without the intervention of anything mechanistic, leaving universal natural history free to embed this Neoplatonic connotation as it seeks to avoid mechanistic "prime mover" causal regress cosmologies; and secondly, the utility of a supposed universal natural history for performing a scientific cross-section on the imagined full composition of the universe in the present.

In a sense the new turn to a scientific "cosmology" added a third stage to the naturalist world-construction Jacobi opposed: it requires tearing away everything else to try to make a world using our own terms alone, destroying the metarational connection with the really existing world and denying the insurmountable limits inherent in using maps of necessity-relations and causal reason as an approach to what really exists; it

[70] Beiser, Paradox, 223.

[71] Schelling, *First Outline*, 6.

[72] This also reflects the Herderian influence on German romantic ideas of nature, especially the influence of Herder's Outlines of a Philosophy of the History of Man (Ideen zur Philosophie der Geschichte der Menschheit) (1784–91), in which he intertwines universal, solar-system, and planetary-wide natural history with human history, strongly evoking ideas of organic wholeness at work in nature and blending the material and immaterial (and mind-like) to evoke a natural cosmos which both evades our full understanding and provides tempting glimpses of its inner structure: "From stones to crystals, from crystals to metals, from these to plants, from plants to brutes, and from brutes to man, we have seen the form of organization ascend … Here the series stops … he seems the highest point attainable by terrestrial organization. … The more elaborate the organization of a creature is, the more its structure is compounded from the inferior kingdoms. This complexedness begins underneath the earth, and grows up through plants and animals to the most complicated of all creatures, man. … Thus there prevails in the invisible realm of creation, not only a connected chain, but an ascending series of powers"; "Nothing in nature stands still: every thing exerts itself, and pushes on"; nature contains a "series of powers urging onward … progressively unfolding themselves." J. G. Herder, Outlines of a Philosophy of the History of Man (Ideen zur Philosophie der Geschichte der Menschheit), trans. T. Churchill, New York: Bergman Publishers, London 1800/New York 1966 [1784–91], 107–108; 114.

incorporates the German romantic response to Jacobi which sought to ennoble this exercise as the finite human mind seeking communion with an infinite it is convinced it is able to intuit and reliably access because of its ultimate unity with it; and thirdly, now, it also moves into using a newly concretizing and exact language for the first two steps by interpreting them as tasks suited to treatment by exact science and mathematics. Spinoza himself, again, would not have pursued the same claims.[73] The early German romantic view of Spinoza's Nature-nature had returned in Spinoza's stead here, too, just as Lazier points out (per above) was the case on the theological and philosophical side of Spinozism's return in the interwar Spinoza Controversy "configuration." As Balashov notes, Einstein "set a notable precedent" for connecting "the material structure of the Universe and its nomic structure[74]"; he also set the precedent for [romantically] interpreting a scientific model as if our epistemological limits – the nomic requirements we have as humans – had not in any regrettably limiting way, but only in a valid and necessity-respecting way, pre-culled what the universe could "be." As just noted, Einstein's 1917 approach overlapped with, and was quickly added to and tested and refined by, the cosmological approaches of others, including in light of Hubble's astronomical observations which added perceived universal expansion and thus a finite temporal past historical duration for our observable universe to the model. Georges Lemaître was among the first to resolve Einstein's models reconfigured[75] as describing an expanding observable universe (which "must" therefore have a finite past history) following the detection of apparent cosmological redshift by Hubble in the 1920s.[76] Lemaître did begin the universe from a single point, but [neo-]romantically still wished

[73] Spinoza, again, must be distinguished from Spinozism, as I have endeavored to make clear throughout this work. Melamed notes that "Spinoza rejected Descartes' view that mathematics deals with the structure of nature itself," cf. Yitzhak Y. Melamed, 2000, On the Exact Science of Nonbeings: Spinoza's View of Mathematics, *Iyyun: The Jerusalem Philosophical Quarterly,* 49: 18. For Douglas, "Spinoza is neither an 'ontological' nor a 'methodological' naturalist" in the contemporary senses of the terms; Spinoza did not hold a "commitment to natural science as the best or only way to learn the truth about the natural world", for example; cf. Douglas, Was Spinoza, 78 and 82.

[74] Yuri Balashov, 2009, A cognizable universe: Transcendental arguments in physical cosmology, in *Constituting Objectivity,* eds. M. Bitbol, et al., Dordrecht: Springer Netherlands, 270, 273.

[75] (Along Russian mathematical cosmologist Alexander Friedmann's route.)

[76] Cf. e.g. Helge Kragh, 1990, *Dirac: A scientific biography,* Cambridge: Cambridge University Press, 224.

to avoid any appearance of deistic mechanism, embracing, again, a Schellingian inverted Neoplatonism in his conception of beginnings in nature as open and simple, as well as lawlike, to be followed by the emergence of increasing complexity. He was arguably also working within the romantic framework of intuiting and revealing already-omnipresent nature using reason – the throughline of scientific "cosmology" ever since 1917, as we will see – which was what actually obviated the need to discuss its being as such (which for Lemaître, a theist, would bear on its status as the creation of God) and enabled the mere in-principle observability of certain laws of nature holding true to function as a shorthand for an entire universe where this hypothetical holding-true took place and in so doing determined, in some quasi-ideal, quasi-real way, the world we observe around us in its entirety (more on this below). Though he did postulate a kind of "primordial atom", "the atomic weight of which is the total mass of the universe", he was adamant that his primordial atom was not the same as, or linked with, the Genesis creation moment.[77]

The new mythology of reason-like self-interpretive aesthetic which characterized parts of the modernism of the logical empiricists, per above, also arose in the new physics and cosmology. Evoking the romantic dream of making all of human existence into integrate-able (and derivable) infinitesimals of the single unity,[78] in 1918, Einstein remarked that while the physicist must for now "content himself with describing the most simple events which can be brought within the domain of our experience" since "all events of a more complex order are beyond the power of the human intellect to reconstruct with the subtle accuracy and logical perfection which the theoretical physicist demands", this is not a permanent state of affairs.[79] Though the physicist *seems* to only "know such a tiny section of nature thoroughly", he is in contact with the whole of it, such that one day, "it ought to be possible to arrive at the description, that is to say, the theory, of every natural process, including life, by means of pure deduction, if that process of deduction were not far beyond the capacity of the human intellect[80]". Einstein's contemporary, mathematician and idealist

[77] Georges Lemaître, 1931, The beginning of the world from the point of view of quantum theory, *Nature,* 127: 706.

[78] Cf. Novalis' view that, as noted in Chap. 1, "it must be possible to make even "the individual moment, … the individual situation etc." universal and absolutized"; cf. Chap. 1, and Novalis, *Notes,* 14 (no. 87).

[79] Einstein, Principles of Research, 43–44.

[80] Einstein, Principles of Research, 44.

cosmologist P. A. M. Dirac, also held that mathematical physics was the key to finding a description of the whole of nature such as could incorporate human existence into its knowledge. He remarked in 1939 that any "limitation in the extent to which mathematical theory applies to a description of the physical universe … is so unsatisfactory that I think it safe to predict it will disappear in the future[81]". He hoped for the future to find "a scheme in which the whole of the description of the universe has its mathematical counterpart" such that "a person with a complete knowledge of mathematics could deduce, not only astronomical data, but also all the historical events that take place in the world, even the most trivial ones[82]". Pioneering cosmologist A. Eddington discussed ways in which Nature or "World" reveals "true natural laws inherent in the external world" which prevent the universe from being "built from elements which are purely chaotic" and instead ensure it is "built from" atoms and point-events and other intelligible, patterning realities, even to the point of leaving us able to "distinguish the actual order of nature from other conceivable conditions" (a very Schellingian, Spinozist set-of-all-possibles-based claim) and able to find "the true laws of nature[83]".

The interpretation of the new cosmology as a realist account of our entire observable universe did still meet with resistance from some less romantic scientists and philosophers in the era. Jeans, himself a pioneering cosmologist, wrote that while the "secret of nature has yielded to the mathematical line of attack" it comes at the price of requiring we also hold that "nothing seems to possess any reality different from that of a mere mental concept", that we cannot even "attribute any reality to the space of the universe, except as a mental concept"; that even the new spacetime "continuum" we must also treat "as a mere mental concept", that the distinction between realism and idealism may have been rendered obsolete, and that if there remained one at all, he sided with idealism.[84] Milne, who first coined the term cosmological principle (cf. Chap. 4), was also a self-stated idealist (and theist) about the meaning of the new science. Philosopher of science Meyerson accused cosmologist and mathematician H. Weyl of repeating the mistake of the German romanticism- and

[81] P. A. M. Dirac, 2003, The relation between mathematics and physics, *Resonance*, 109.

[82] Dirac, The relation, 109.

[83] A. S. Eddington, 1920, The meaning of matter and the laws of nature according to the theory of relativity; Mind 29: 156–157.

[84] James Jeans, 1932, The Mathematical Aspect of the Universe, *Philosophy*, 7: 12–14.

Schelling-inspired nineteenth century *Naturphilosophien*, by "maintaining[, like them,] that nature is completely intelligible" and abolishing "the thing-in-itself" leading to the "identity of self and nonself[85]". In a broadside in [appropriately enough] the journal *Nature* in the 1930s, one well-known astronomer, Herbert Dingle, per above, opposed the new cosmology as unwarranted wish-fulfillment.[86] In a sense partly evoking Jacobi, Dingle faults the extension of the "scientific sanction which had been earned only by the underlying system of mechanics" to its utility as a system of cosmology.[87] He calls this a "mistake" and expresses his bewilderment that scientists would claim relativity theory as such could tell us something about "the peculiar character of any particular system[88]".

Despite these critiques cosmology was undeterred. Nothing could break into the new monistic nature from without; it barred all other speech of the universe.

Barth's Negation of Spinozism

In the first Controversy, Jacobi was essentially arguing that theoretical reason and its necessity-based constructs must be understood as functional within, but Other to, the external, really existing world, else nihilism, including self-God identity, follows. Similarly, this Weimar Era return of the Controversy "configuration" saw, ranged against all the resurgent currents of neo-Spinozism and neo-German romanticism just sketched, the "crisis theology" of Barth and his allies (per Lazier, primarily Gogarten and Thurneysen) arguing that theoretical reason and its world, i.e., the reason-amenable aspects of nature (which they were willing to grant more objective reality to than Jacobi did, but assigned them even less final importance than he), must be understood as unable to grasp or contain the Real, i.e., God, who must be understood as radically Other, else nihilism, including self-God identity, follows. And while Barth and his colleagues were perhaps most deliberately focused on countering their era's upwelling of liberal, romantic, Schleiermachian theological exuberance, as

[85] Ryckman, *Reign*, 239, citing the arguments against Weyl levied by 1920s Polish philosopher of science Émile Meyerson, including in Meyerson's *La Déduction Relativiste* of 1925.

[86] Dingle, Aristotelianism, 785. This echoes Halmi's attribution to the early German romantics of "theoretical wish-fulfilment", cf. Halmi, *Genealogy*, 19.

[87] Dingle, Aristotelianism, 785.

[88] Dingle, Aristotelianism, 785.

noted by Lazier,[89] their views also hold oppositional force when placed across from the philosophical and scientific forms of neo-romanticism we have just been discussing, particularly if and as we make an effort to draw out some of the implicit environmental dimensions of the Controversy's return.

While Barth offered a more directly theological version of Jacobi's insistence on external reality as Other to theory (thus also making theory other to it), he also echoed some of Jacobi's emphasis on the peril involved with claiming to be able to use theoretical reason to plumb the true extent of what was real. Jacobi's position was also partly reprised, though again with a difference, through Barth's de facto use of, and insistence on the high and even existential stakes of preserving, the metarational faculty or capacity within the subject.[90] In this period, as just discussed, much of Western (especially Germanophone and Anglophone) philosophy was beginning to be fully subsumed by scientific and mathematical questions, particularly those arising from the new physics and its new monism, and this science itself was deeply imbued with neo-romantic and neo-Spinozistic views of its exclusive true rational access to nature and reality. This in a sense made theology potentially the last remaining hospitable territory for the Jacobian metarational subject opposed to theoretical world-construction, although even here, Barthian theology was also having to carve out space apart from the prominent liberal neo-romantic theological approaches surrounding him.

Barth and aligned thinkers such as Thurneysen were intent on doing just this; they opposed, per Lazier, the liberals' rationalized, naturalized approach to reality and God as it tainted God with "worldly corruptions," even going so far in the other direction that for Lazier, they resembled the ancient gnostics.[91] Morally and theologically, "[i]n order to save God,

[89] Lazier, *God Interrupted*, 5–6.

[90] Again, "subject" meant in the previous chapters' Jacobian sense as referring to the full person or the whole inner self, as evoked for example in his notable passage in *Letter to Fichte* where Jacobi writes, "I declare that my reason, my whole inner self, flies into a rage before a representation of this sort, it shudders in horror and fright; I recoil from it," regarding the "weaving of weaving" creation of reason being substituted for the real external world, cf. above and cf. Jacobi, Letter to Fichte, 511 (no. 23).

[91] In his Preface to the 2nd edition, Barth himself notes that his resemblance to the ancient gnostic thinker Marcion is only surface; cf. Karl Barth, 1933, *The Epistle to the Romans*, trans. Edwyn C. Hoskyns, Oxford: Oxford University Press, 13.

they banished him from a world soaked in sin;[92]" intellectually, they removed God from a world soaked in Spinozistic enthusiasms ready to distort who or what God actually was and offer easy recipes for how God might be directly and comfortably known. Barth also had some Jewish theologian and philosopher analogues,[93] who were likewise seeking to carry out a "partial retrieval of nonrational elements in the Jewish tradition" in contradistinction to the new Spinozism (and the old Enlightenment, Mendelssohnian paradigm of social and religious integration and reasoned "synthesis" with Protestant Europe which had continued to feature prominently in European Jewish culture through the 1800s).[94] One of these Jewish scholars, Strauss, wrote his dissertation on Jacobi and used this "resurrection" of the Controversy to find the "conceptual tools by which to put [liberal theology] down[95]". Strauss found echoes of Jacobi in Rudolf Otto, a Marburg theologian also inspired by another philosopher named Jakob Friedrich Fries; Otto wrote the "most widely read work of theology in interwar Germany," *Das Heilige* (The Idea of the Holy).[96] In *Das Heilige,* Otto writes of how the true manifestation of the holy is found not in representations of it as spirit (or any other positive or mythical representation), but rather in "a peculiar 'moment' of consciousness," in which the mind experiences a "*stupor* before something 'wholly other'" (emphasis original).[97] Here, and in Barth, the Other becomes even more focused on the divine alone, while for Jacobi Otherness arguably implicitly extends to the entire external world in the sense of its real existence and being. In both Barth and Jacobi, however, registering the world (or God) as Other to our selves and to the grasp of theoretical, necessity-based

[92] Lazier, *God Interrupted,* 6.

[93] Lazier aligns with Barth, on the Jewish theological side, F. Rosenzweig, L. Strauss, and Strauss's earlier influence, the nineteenth century thinker H. Cohen; Rosenzweig and Strauss assessed the excommunication of Spinoza as valid, in contrast with the majority view among Jews in the Weimar era; cf. Lazier, *God Interrupted,* 85f. Lazier also indirectly links D. Baumgardt to the anti-Spinozan side, noting that Baumgardt "resurrected [Jacobi] … as a harbinger of the spirit of the age," calling Jacobi in 1933 the "philosophical founder of an ethical standpoint that had resurfaced in his time with the 'widest currency'", linking him with Kierkegaard – who had in turn inspired the crisis theologians as well, cf. Lazier, *God Interrupted,* 98.

[94] Lazier, *God Interrupted,* 6–7, quotation from 7.

[95] Lazier, *God Interrupted,* 94.

[96] Lazier, *God Interrupted,* 94–95.

[97] Rudolph Otto, 1926, *The Idea of the Holy,* trans. John W. Harvey, London: Humphrey Milford, 27.

abstraction is the crucial feature of its Otherness, and the connotation of something or someone as Other differs sharply from how the notion of encountering an entity or a person as the "Other" would later carry negative connotations in certain disciplines.[98]

While Otto's work drew significant attention, in Lazier's view Barth's *Der Römerbrief* (*The Epistle to the Romans*) "set the terms – whether embraced or disputed – for much of Christian and Jewish thought in the wake of Versailles".[99] *Epistle* "hit the intellectual landscape [of post-WWI Europe] like a bombshell", "depict[ing] the general human situation as one of profound crisis and only in light of this overall negativity suggested hope[100]". In the *Epistle*, in the guise, and real pursuit, of offering exegesis on the Biblical text of the Book of Romans, Barth takes a hammer to liberal, rationalized notions of God's immanence in history and God's amenability to human forms of experience, expression, and causal reason. His main response to the returned neo-Spinozistic and neo-romantic, pantheist and monist, stance toward nature, God, and reality is, simply, "*Nien*" *[No]*: No to placing any great significance in such a world; No to thinking reality is easy to penetrate using monistic theoretical reason; No to thinking positive and rational knowledge of the divine is possible. Only a negative theology was possible, one evoking Jacobi's "non-philosophy" and its intentional self-placement outside the pale of mainstream naturalism's assumptions about how to legitimately ontologize:

> Those who take upon them the divine 'No' shall themselves be borne by the greater divine 'Yes'. ... It is the faithfulness of God which we encounter so unescapably in the prophet's 'No': God the Holy One, the altogether Other. It is the faith of men which we meet in the awe of those who affirm the 'No' and are ready to accept the void and to move and tarry in negation.[101]

[98] Especially in fields with strong German romantic historical influences such as psychology, sociology, and anthropology – fields where the romantic type of Othering had to eventually expose its true meaning as monistic conquering or altering subjugation, since in the romantic worldview there is arguably no positive or genuinely incommensurable type of Other permitted to exist, there is only the "I" and its monistic internalism. Cf. also Chap. 2.

[99] Lazier, *God Interrupted*, 23. Here Lazier is citing also Mark Mazower, 1998, *Dark Continent: Europe's Twentieth Century*, New York: Knopf, 3–40.

[100] Sigurd Baark and Christine Svinth-Vaerge Põder, 2023, Crisis and Reorientation: Introduction, in *Crisis and Reorientation: Karl Barth's Römerbrief in the cultural and intellectual context of post WWI Europe*, eds. Christine Svinth-Vaerge Põder and Sigurd Baark, Cham: Palgrave Macmillan/Springer, 2.

[101] Barth, *Epistle*, 41–42.

Barth holds that "The Gospel proclaims a God utterly distinct from men" and as a narrative account constitutes a different, Other, kind of true thing, "the clear and objective perception of what eye hath not seen nor ear heard[102]". The realm of the utterly transcendent God touches the earthly realm only in a single tangent point,[103] the non-analyzable appearance of God to man. There is always and forever only this single point [or typology] of contact between God and our world; God is the tangent-line of Day, in Barth's metaphor, to our corrupted circular existence dwelling in Night. God is the "unknown God," and knowledge of the "appointment of Jesus to be the Christ takes place in the Spirit and must be apprehended in the Spirit. It is self-sufficient, unlimited, and in itself true[104]". There is no standard against which to evaluate such a claim. What in this context we might call Barthian metarational cognition is the only way the human can grasp or think it, while also protecting it from being falsely subjected to rationalistic proofs or tests, and protecting God from being "confounded with any high, exalted force, known or knowable[105]".

To make room for God's undeniability while preserving God's radical distinction from what we commonly think of as truth, humans must approach God negatively and "know" God "only as contradiction[106]"; humans must wait, and make a "clearing", indicating, negatively, the other realm, the realm of God.[107] This clearing, and Barth's overall emphasis on a negative theology, evoke, again, some of Jacobi's efforts to clear a space in which the majority might be found wrong in their conception of what generates true [or ultimate] knowledge. If we could extend Barth's notion of making a clearing for a realm beyond human necessity-mapping theoretical reason, into a spatial or environmental key, we might find a normatively positive link to valuing wilderness (or a truly clear night sky, devoid of light pollution) perceived as a symbol of moving away from human reason and its (trivial, for Barth) certainties; or a normatively negative link, where the spatial correlate of the "clearing" and "waiting" is accepting the inevitability of a human-levelled, destroyed world from which we have cleared all life and from which only God, for Barth, can save the human being.

[102] Barth, *Epistle,* 28.
[103] Barth, *Epistle,* 29–30.
[104] Barth, *Epistle,* 36.
[105] Barth, *Epistle,* 36.
[106] Barth, *Epistle,* 38.
[107] Barth, *Epistle,* 39–40.

Paralleling and echoing Jacobi's critique of humans replacing the real world (and God) with a *nihil* that leads to their self-enthronement as God (cf. Chap. 2), Barth faults humanity's worship of a "No-God" which is really only itself. This self-worship also turns out to entail a false "super-world" being assumed, one amenable to our reason; this "super-world" resembles a Spinozistic nature, and Barth's reaction to it echoes Jacobi's, for example where Barth writes:

> ... we assume that we are able to arrange our relation to Him as we arrange our other relationships ... [and] press ourselves into proximity to him ... We dare to deck ourselves out as His companions, patrons, advisers, and commissioners ... Secretly we are ourselves the masters in this relationship ... We are not concerned with God, but with our own requirements, to which God must adjust Himself, ... [while o]ur arrogance demands that, in addition to everything else, some super-world should also be known and accessible to us ... Our devotion consists in a solemn affirmation of ourselves and the world and in a pious setting aside of the contradiction ... what is called 'God' is in fact Man.[108]

For Barth, God must not come to us via reason but must break in upon us "through the – *clearly seen* – memory of eternity" (something of a Jacobian and a Platonic notion), "breaking in upon our minds and hearts," and if this "'breaking in' does not occur, our thought remains merely empty, formal, critical, and ... incapable of mastering the rich world of appearance"; "Unbroken thought thereby divests itself of any true relation to the concrete world[109]".

Only broken thought is a legitimate form of knowing reality; this, we can recall, resembles the Jacobian position that the first thing we encounter, before thought, is the Thou, which announces itself to us from without, shaping our reason itself before we can even conceive of ourselves as an "I." Barth forms a striking contrast with the Vienna Circle and related approaches where unbroken, logically connected and ostensibly self-supporting thinking is prized as the only legitimate speech of the real structure of the world. Barth directly faults the new cosmological pretensions of the sciences, saying they give the subject "a COSMOS which, because we no longer know God, is not Creation[110]". In Baark and Põder's

[108] Barth, *Epistle,* 44.
[109] Barth, *Epistle,* 48.
[110] Barth, *Epistle,* 168–169.

view, Barth objects to the fact that humans are unable to "transcend time-bound human finitude," yet they "posit theoretical constructs beyond the deconstructive reach of time and mortality", again suggesting the grounds of his opposition to the burgeoning scientific, neo-Spinozist, monist worldview, here on simple epistemological grounds.[111]

The notion of a "broken-in-upon" human being – the human who is seized by some form of contact with a wholly external and Other divine realm which shatters everyday norms, and attains a higher state of awareness as a result – is an ancient mystical notion, one which well predates and exceeds Christian thought. As, e.g., Feldt's research suggests, belief in a need for the human to be broken in upon to reach proper levels of awareness can also (though not always) carry deep-running environmental correlates which challenge the human's right to dominate and "use" the non-human in unthinking ways, since cultivating and sustaining the possibility for this breaking-in to repeatedly take place can require certain spatial and tangible practices including the preservation of sacred lands or places (set apart from everyday profane ones) seen as both symbolic and physical gateways to the supra-mundane influx being sought.[112] The non-human world, especially barren or remote wilderness, stood for Otherness and ultimately the juncture point between the human and the divine in many past societies for thousands of years, before the onset of the type of Western monist societies which ended these beliefs in large part (though their traces persist), and there were often, again, some de facto preservation practices (like tabooing the clearing of sacred groves or the regular inhabiting of sacred zones) associated with these perceptions of it.[113]

In Barth's case, Lazier notes specifically that Barth's views seem to threaten "systems of knowledge that … [conceive] of the world as an enclosure and of themselves as sovereign within it[114]". Barth characterizes all forms of unbroken, rational thinking as belonging to the "Night" which leads only to "Death[115]"; though he is speaking in the soteriological sense, given the enormous amount of literal death on earth (both of other humans and of all other species) humans have caused while convinced that

[111] Baark and Põder, Introduction, 6–7.

[112] Cf. e.g., Laura Feldt, 2012, Wilderness in Mythology and Religion, in *Wilderness In Mythology and Religion: Approaching Religious Spatialities, Cosmologies, and Ideas of Wild Nature*, ed. Laura Feldt, London: Equinox, 1–23.

[113] Cf. e.g., Feldt, Wilderness.

[114] Lazier, *God Interrupted*, 23.

[115] E.g., Barth, *Epistle*, 54.

their own constructions really exist, and that they have the right to make them real by force as a means of "uncovering" their reality (and also to instantiate other smaller "worlds" built solely using relations of necessity and thus through force, too, as we will discuss in the last chapters), Barth implies there might be a different spatial and physical (and thus environmental) correlate to his own position which would be more aligned with "Day" and "Life". Even though Barth himself does not prioritize what happens here on earth relatively speaking, this environmental echo seems to be present in his metaphor.

Arguably as far back as Plato's co-option of the cave into a metaphor indicating the need to leave the earthly to pursue true knowledge (contrasting with the traditional religious use of caves as places of enlightenment), the West has arguably lacked a truly rural philosophy – a kind of philosophy which indeed might go a long way toward filling in the outlines of Jacobian environmental cognition we are trying to delineate in this book.[116] But the nineteenth and twentieth centuries' industrial production changed the tools available to humans in acting out their anthropocentric beliefs and building their rationalized worlds, and the growing human population changed the scale, scope, and speed with which these beliefs and worlds could be acted upon and built. While Barth would not necessarily have counted the anthropogenic erosion of the natural world's integrity or our failure to value nature as a key part of humanity's state of crisis, given the blanket quality of his diagnosis of the fallenness and ignorance of the West of his time, it is not too much of a stretch to include its environmental behaviors as those which might also stand to be changed, perhaps (though not necessarily) in normatively positive ways, if it were to allow itself to, as Barth urges, be broken in upon by the actually Other once more.

Barth himself was only intent on arguing the conditions bearing on human spiritual survival and ultimately salvation. For Barth, this depends on letting ourselves be broken in upon by what is Other, and this is only God. Yet Jacobi's more world-affirming call for the recognition of the need for the breaking-in by the external Thou and Other upon the I is not so distant. Both men are making at root the same claims: the least-negotiable, highest form of truth cannot be a product of our theoretical

[116] The West's 2500-year lack of a true rural philosophy (or a philosophy of nature which is really externalist) as opposed to an urban one focused on human needs and the order of the *polis* is the subject of my next work, *Rural Philosophy,* in progress.

reason; our theories cannot touch what actually makes being; since we are nonetheless still beings ourselves, our most real condition can only be seen through the use of that metarational faculty which is really part of us, really "of" our being. Barth's saying-No to our earthly realm does still also have, again, and as Lazier notes, tinges of gnosticism, and to the extent we take him to really be opposing the earthly sphere as a whole and not just the making-ultimate of a rationalized or "enclosure"-style (in Lazier's term, cf. above) version of it, this may, again, point to a less normatively positive environmental reading of his approach compared with the Jacobian position. Jacobi's position is more conducive to environmental cognition, while Barth is trying to orient people toward clearing away everything about this life in order to receive God's grace, which for Barth so far exceeds our ability to seek it rationally that when we try, we end up sliding into a state of self-worship. Still, if Jacobi seems to afford the subject more space than Barth to apply the fruits here on earth of radically rethinking naturalism, Barth's skepticism toward unbroken thinking and the cognitive errors of [neo-Spinozist] naturalism which lead to self-worship may still bear some of the same environmental interpretation.

Barth also echoes Jacobi's prescience regarding the nihilism of a world in which only ontologies of necessity manipulable by us were "believed" in. Barth writes that if human [theoretical] reason fails to understand its true limits, "reason itself becomes irrational[117]"; this echoes one of the arguments which Jacobi tried in multiple ways to present. The capacity of reason itself to become irrational was front of mind for many in Barth's era because of the horrors of the First World War; it was even a preoccupation of the German speaking (and Anglophone) intellectual class whether and how German philosophy itself may have fueled German militarism and the war's outbreak.[118] This was, in different ways, acknowledged (or claimed) by some Germans themselves, among them the H. Cohen noted above as an influence on some of Barth's Jewish analogue irrationalists.[119]

Part of what was at stake in the original Spinoza Controversy was, in Beiser's view, a debate over why human subjects should be "loyal to reason

[117] Barth, *Epistle*, 53.

[118] Andreas Vrahimis, 2015, Legacies of German idealism: From the great war to the analytic/continental divide, *Parrhesia*, 24: 83–106.

[119] Regarding H. Cohen, cf. Footnote 93, above. Re the intricate and partly political historical question regarding the role of German philosophy in WWI cf., e.g., Vrahimis, Legacies.

[as opposed to revelation] if it pushes us into the abyss[120]" of war, revolution, and social upheaval. We can say this fear of reason leading us into an abyss is a defining characteristic of the Controversy each time it recurs: in our own, third iteration we are confronted with the brutal destruction of much of the planet's biosphere, continued gratuitous wars in the shadow of nuclear weapons, and the proliferation of separate spheres of subjective rationality (along with a rising threat of the undermining of all sense-making by a new belief in the sheerly informatic view of the world) as our own challenge to reason itself. But casting Jacobi or his successors as irrationalist, as Beiser does in *The Fate of Reason* (1987), can be seen as slightly disingenuous – or, put another way, as only possible because the Spinozist side has both times triumphed. In the era of this second Controversy "configuration", Europe had just been dealt another round of brutal historical evidence of reason's capacity to lead enlightened nations into Beiser's aforementioned "abyss"; was it more rational to then double down on ontologies of necessity in the war's aftermath, or, like Barth, instead diagnose humanity as now occupying the kind of state of crisis in which humans could only sit and wait – wait for a higher kind of truth to return to them from without? Jacobi and Barth oppose theory with metarational cognition, not with irrationalism in any general sense. Their position is irrationalist only with direct respect to being opposed to basing our existence, and our understanding of reality as a whole and in the ultimate sense, upon a rationalism of mere necessity- and logic-construed webs. It is metarational in terms of its overarching *cri de coeur* for the subject to return to a higher and better use of its reason as a whole, including using it to recognize why and how we must circumscribe our belief in and comfort with these types of webs, and avoid using them as deceptive stand-ins for the world and its actual external truths, truths less comforting but of far more importance. Barth's argument that we could see that the deployment of unchecked, uncontextualized, un-"broken" causal reason as our only mode of understanding reality was itself irrational implied that he believed, like Jacobi, in the metarational, full, subject, and effectively assigned it the task of saving the human being by reorienting and reopening it in a radical way. Its first task was to avoid leaving the Western subject in a state of crisis wherein the human lacked even the single saving, piercing, divine tangent line offering potential redemption; a state of crisis

[120] Frederick Beiser, 1987, *The Fate of Reason: German philosophy from Kant to Fichte*, Cambridge: Harvard University Press, 80.

related to, though not identical with, the slide into nihilism with which Jacobi associated naturalism.

Barth used a new version of dialectics in his positions, again, not eschewing reason as such for irrationalism, but instead arguing for and trying to establish something like a metarational understanding of our true condition as human thinkers and would-be knowers. As Keedus points out, the crisis theologians intentionally opposed the Hegelian (and, we can add, German romantic) notion that dialectic leads to a synthesis of opposites, and the Schellingian belief in "'the totality of existence as one… mind-subject'".[121] For Keedus, while "Hegelian dialectic results in unity and coherence, theological dialectics [in the Barthian vein], though also a process of [a] thinking together of contradictory things, has the opposite goal of demonstrating profound difference, thus revealing the core of each entity in the equation[122]". Hegel's dialectic reveals itself to be undergirded by a Schellingian assertion of unity and monism, and we can link this to Jacobi's own observations of the reducibility of all logical necessity-claims to mere identity-statements. Barth wanted to keep the approach to the Gospel from falling into these meaninglessly circular logical exercises; he innovates with his negative or crisis dialectics in a way which again evokes Jacobi's desire to carve out space for a different kind of knowledge that was not formally like other types. For Barth, in Baark and Põder's view, the "universal and unsublatable negation encountered in the early textual witnesses to Jesus of Nazareth as the Christ" offered a different "form of practical certainty," not theoretical certainty or abstract proof, "in the sovereignty of God[123]". True incommensurables, and the question of our [in]ability to recognize and preserve them, also have potential spatial consequences: the world we live in today, in the twenty-first century, shows little tolerance for treating any spaces or species as incommensurable with being subjected to and subsumed by human aims or "worlds". Seeking to preserve the natural world in itself as incommensurable with the solely human, except as the human is prepared to experience it as a Thou, is one

[121] Liisi Keedus, 2023, 'As a Tangent Touching a Circle': Karl Barth and dialectical theologians rethinking time after 1918, in *Crisis and Reorientation: Karl Barth's Römerbrief*, 44–46; on 44 citing Brunner in Emil Brunner, 1929, *Theology of Crisis*, New York: Scribner, 95.

[122] Keedus, 'As a Tangent,' 46; also citing Emil Brunner, 1981, *Die Offenbarung als Grund und Gegenstand der theologie*, in *Ein Offenes Wort I: Vortrage und Aufsatze 1917–1962*, ed. Rudolf Wehrli, Zurich: Theologischer Verlag, 113–114.

[123] Baark and Põder, Introduction, 6.

potential environmental extension or interpretation of Barthian crisis dialectics.

Overall, in the crisis theology position, despite it carrying a more explicitly theistic message than Jacobi's philosophical approaches, we can find an echo of Jacobi's basic requirement that we preserve ourselves as the kinds of subjects which can know and experience true exteriority and understand what it means for human causal reason's pretensions to world-make; and can actually be *capable* of meeting (should we choose) the requirement that we approach the really-existent world with a willingness to be broken in upon by the truly Other. Thurneysen wrote that one of Dostoevsky's insights was that there is an "absolute enigma that lies over existence" that suggests "the meaning of life is not to be found in this life[124]". But just as Jacobi arguably meant by "supernatural" first and foremost that which was super- or beyond "the complex that we call *nature*[125]," so too it is possible to find in the crisis theologians an opening onto deriving meaning from, and seeing as inestimable, a non-human world re-cast as not "in this life" simply by being reconsidered out of the shadow of the claim which Lazier notes Barth's position finds problematic and even nonsensical, namely, the [neo-Spinozist, monistic] claim to have rendered the world "an enclosure" with ourselves "sovereign within it[126]". Once unenclosed cognitively (and potentially also physically, as an environmental correlate of this shift), the non-human world could once more prove able to form part of the meaning of life, including through intertwining with the subject's ongoing task of ending its misguided notion of ultimate sovereignty.

Crisis Theology Versus Critical Monism as Crux of the Second Controversy

Having almost arrived at our third, contemporary iteration of the Spinoza Controversy, I will offer here a change in nomenclature from continually writing neo-Spinozistic and neo-German romantic, to calling the increasingly dominant and mainstream anti-Jacobian position critical monism, or critical monist in the adjectival form; for both cases I will use the

[124] Eduard Thurneysen, 1964, *Dostoevsky: A theological study,* trans. Keith R. Crim, London: Epworth Press, 46.

[125] Jacobi, Concerning the Doctrine, 376.

[126] Lazier, God Interrupted, 23.

abbreviation CM. CM builds a claimed ontology of the entire presumed-infinite universe sheerly from statements of logical necessity, identitarian in essence, whether these in turn are based upon only more pure mathematics and logic, or upon demonstrated laws and facts first observed in nature and tested empirically and then converted into the language of mathematics and logic (and the natural sciences), like the laws of gravity or atomic behavior. In CM, as we have seen, what makes these collections of observations able to signify a monist whole (and a claimed exclusive standard for meaningful speech regarding nature) is an often tacit, sometimes overt, reliance on romantic and Spinozist assumptions about the world really being, and only being, a real-ideal quadratic type of [Nature-]nature (cf. Chap. 2). The CM universe is known and constructed through theoretical reason, and mathematical logic, but also declared to be unified with all other fields of knowledge, and to have no inherent distinction from, or incommensurableness with, the sense of beauty or artistic meaning. Despite claiming "critical" awareness as to the possible inherent limits of positive human knowledge, this is mostly an aesthetic maneuver to prevent any type of Kantian critique from undermining the resulting "nature" of the worldview. With time, and devoid of competition since no alternative to positive scientific claims about the infinite all are permitted to count as knowledge of nature, any actual Kantian self-awareness that might have been present in principle in the German romantics' view will now, as of the end of the second Controversy iteration, begin to fully vanish. Emerging out of the interwar period, and moving into post-World War II discourse, CM's nature is more than ever only a malleable and directly knowable quadratic. CM excludes non-necessitarian, non-CM speech of nature as not of the proper category, and, again, conceives of all human knowledge as centered in and underpinned by the theoretical capacity, particularly by mathematical physics. By the early twenty-first century, as we will see, CM still effectively claims to be an absolutized (in the romantic sense) knowledge-system regarding nature, but it has also spawned many smaller-scale imitators of its world-building approach. CM perceives its view as realist, such that the kinds of necessity relations it presents as the ontological tendons of the world really derive from, and directly symbolize, an external, objective state of affairs, *and* also as idealist, since these relations are perfectly intelligible (as laws or constants) and are also seen as actually generative of the universe itself. At the same time, as we have discussed, this obscure (or even nonsensical) threshold for the existence of nature is not the final one in CM; CM also utilizes (and could not remain even passably coherent without utilizing) a pre-existing and infinite Nature just "off-stage" from the models and theories of the universe themselves, a Nature which grounds cosmological or master-ontological theories and findings, but does not mechanistically interfere in them. This Nature is one which has all the neo-Spinozistic and romantic a prioris built in: eternity and infinity (albeit only in the sense of making temporality and finitude perfectly generic or categorical); assumed productive power; alignment with a set of all possibles; and ultimate identification sheerly with the concept of necessity itself – its most direct hinge to finite and observable nature, with which it ostensibly shares this key essence.

In CM usage, Nature and nature can [solely] exist in the form of any of the three types of provable necessity-relation we have discussed (logical-mathematical, rational-speculative, or artistic [now increasingly less artistic, and more technological-production-based]). At the same time, the monism aspect of CM – displayed in its designation of non-necessitarian truths about nature as categorically impossible; its assertion of a single universal type of "being" (which must come as the result of, or "as", a necessity-relation); and its belief in a single correct method for knowing the real, rooted in the theoretical faculty and its mastery of logic, mathematics, and science, which must also rightfully displace philosophy, and unify with beauty and art – also implicitly lays claim to include a fourth, substantival, blanket kind of necessity relation in (or "as") its universe, the type with which Spinoza himself originally justified his own monism (cf. Chap. 2). As discussed above, while Spinoza himself may have sought to build his system on this type of necessity, Spinoz*ism*, and the radical German romantic empiricizing and reinventing of the God or Nature entity, unavoidably removed this actually ontic kind of necessity from its world (as Jacobi understood, just as he understood that Spinoza's one necessary substance – which he did admire on its own terms for its architectonic integrity – must lose its grandeur and even its coherence if subjected to reinterpretation or reinvention). None of this, however, prevents the CM worldview from often simply assuming, or claiming *en passant* (as was the case with Einstein), that its scientific universe does presence a necessary God or Nature in Spinoza's own absolute, monism-supporting sense. This assumption seemingly helps paper over the self-reference problem, allowing cosmologists (and other scientists) more leeway to simply move and exchange various relational pieces of an already-existent, already generated, real Nature-nature according to logical, speculative, or artistic (or technological) necessity conditions, conditions they can all the more readily actively or outwardly credit with causing the existence of things due to the blanket, substantival necessity, and the absolute ontological reality, they assume, however tacitly (and incorrectly), already saturates their schema.

If the Controversy at its heart pits theoretical reason against metarational cognition, over the question of which of the two has the legitimate ability to curtail and fully characterize or critique the other, CM is a good approximation of where the neo-Spinozistic and neo-German romantic position tends to cluster by the twentieth century, particularly moving toward the later part of the century. The CM view, as a monist one, claims to be self-policing (to the extent it polices or limits its claims at all), rather than needing to be, or even being able to be, restricted or even directly "seen" by something like the metarational faculty. It is loathe to contextualize the claims of theory within some form of higher judgment on their applicability to the real, historical, present world.

The second Spinoza Controversy configuration arose in the form of an irreconcilable discordance between the Barthian theological stand and the assertion of the CM world picture across multiple areas of Western thought.[127] And this configuration, as we have touched on above, did have clear environmental dimensions or correlates. While Barth did not believe in the "religious veneration of nature", diagnosing, as Keedus notes, the "religious veneration of nature and modern culture" together as flawed and errant,[128] Barth's quasi-gnostic rejection of theoretical reason and its power, discussed above, does still implicitly contest all the forms of domi-

[127] Depending on one's interpretation of how the visual arts (including architecture) speak to or construct their viewer, it is arguably the case that many strands of art and architecture in this era, notably including the Bauhaus movement, grouped themselves under the CM banner as well. In at least some of the poetry and fiction of the era, on the other hand, we see the opposite; works like *Disenchantment* (by C. E. Montague, 1922) or *All Quiet on the Western Front* (by E. M. Remarque, 1928), or some of the post-WWI poems of writers like Ezra Pound or T. S. Eliot, seem to preserve a metarational perspective, as they evoke the disorientation and despair the inner self (or whole subject) experienced in and after the war and the problems with thinking of the scientific world view as some type of antidote. We can also note here that Jacobi himself wrote a novel, *Woldemar*, and that Barth's *Epistle* is deeply literary in its form, and perhaps bracket the question of the metarational and even actively anti-theoretical subject emerging via the form and content of literary (versus visual) art during all three Controversy periods as a topic for future exploration.

[128] Keedus, 'As a Tangent,' 40; also quoting Barth in Karl Barth, 1968, Past and Future: Friedrich Naumann and Christoph Blumhardt, in Karl Barth, 1968, *The Beginnings of Dialectic Theology*, ed. J. Robinson, Richmond: John Knox Press, 37.

nation which flow from this form of reason, including all the ways it enforces itself in and over the natural world as well as the human person and the social realm.

Moreover in Barth's poetic description of the *love* of God, of God as the Thou which negates anything and everything that can be generated only out of conditions imposed and controlled by the finite "I" itself, we can also find a link with Jacobian environmental cognition. Love potentially connects the fuller self with the non-human realm as our Other, since love cannot be logically instructed, directed, or tested. The subject experiences it as a kind of pull (or even stupor, per Otto; cf. above), which can only be experienced, and never rationally constructed. It is a pull or connecting ligament which, when it is violated, or broken, creates an emotional, felt response, including and akin to Jacobi's "recoiling in horror" with his "full inner self" when we realize we have lost something real (the real world) by following the *nihils* of our reason too far. Likewise, when Barth evokes what inevitably must follow from our honestly encountering the Other, the extraordinary, terrifying, reckoning which is also somehow at the same time the only balm for our core human condition of fear and suffering, an encounter which demands we immediately reflect, "Who then am I?"[129] in a moment of blinding awareness of our need for radical humility, this same sequence could conceivably occur when considering and encountering the infinite natural world as it is in itself, and as external to reason (a possibility of which the German romantics in particular were well aware, and did also wrestle with to some extent, as noted earlier, though they came down on the side of claiming to keep even the chaotic or sublime infinite still just within reason's grasp).

Another element of Barthian non-theoretical cognition of the truly external real with environmental connotations (not developed by Barth himself) occurs in Barth's view that it is impossible to ask questions such as whence derives the "possibility of the very existence of any biological function[130]" using theoretical analysis. For Barth we must refer this type of question only to God or risk misusing our reason. While Barth might not recognize that the question of the origin of the possibility of the very existence of any biological function, or any other foundational ontological or cosmological question, could also be considered the preserve of something like the non-human infinite itself, while still being barred to theory,

[129] Barth, *Epistle,* 319.
[130] Barth, *Epistle,* 425.

his position does arguably bear this possible environmental reading.[131] Again, as we saw in considering Jacobi's positions on the sameness of the "fabrication" status of humans' rationalized versions of nature and their rationalized version of God (cf. Chap. 1 above), arguably CM's secularism does not only risk asserting the theoretical subject as God, but also as the really real infinite universe beyond us, to which the CM world assigns no inherent being or meaning beyond what is accessible to our theories, thus dictating that no knowledge or truth can be had about it which does not belong to those theories. In a neo-Jacobian position, nature, or non-nature, can regain its self-contained ontic heft as a result of the metarational perspective. Questions which theory cannot pose can return to our awareness and our fuller reason, including the central question of asking from the metarational, external-world acknowledging, perspective, *what* all that really exists actually *is*, what its raw being as such means or contains in itself, and what it means that it is apparently a fully realized physical infinity, something which cannot be accommodated in theoretical reason because as an historical fact, its presence is different from anything which resembles taking a finite category of theoretical reason's own creation and making it perfectly generic, which is all theory can do when it tries to analyze conditions "at" infinity (as we will return to discussing in the final chapters). These, and many others which can return, are potentially all questions for whose answers we can only "wait," as Barth would have his thinking subject do; but they are at least questions we may subject to environmental cognition, and not put to our mere theoretical necessity-mapping capacity (where such questions anyway categorically cannot exist). These are also questions which may bring us toward the moment of radical humility when our fuller reason must reflect, "And what, then, am I?", in the face of the universe in which we physically find ourselves.

Barth's "clearing" and "waiting" states for the subject can thus, in an environmental sense, offer a way of rejecting capitalist value-raising

[131] This takes us again to the question of whether Western "rational" cosmology can ever be rational. Arguably, as touched upon earlier, the global religious and cultural history of human societies before and beyond Judeo-Christian monotheism (and arguably even within this monotheism, in some ways, up until Enlightenment deism) is full of examples of humans effectively using creation myths and other cosmological accounts as ways of framing and curtailing human theoretical (necessitarian) reason itself, and accepting the unavoidable presence of some type of reference point outside its frame, whether mythical, or unacknowledged and thus incoherent – as Jacobi himself relatedly discusses on a smaller scale in his treatment of the illegitimate removal of final causes in Spinoza (cf. above).

actions, engendering contemplation, and permitting non-interfering experience and observation of the non-human realm. In contrast to the transformative quality of romantic knowing, which always identifies and knows through observing change and potential change (and in the post-WWI, and then particularly in the post-WWII, period, increasingly through experiment, mathematically precise measure, logical test, and observer-interference), the world can be perceived metarationally and by the full person, if the person approaches it in a spirit of stillness [akin to the state in which Jacobian spiritual cognition takes place, per Chap. 2].[132]

Barth's seeming valuing of stillness was also linked to his opposing neo-Spinozan neo-romantic interpretations of human history as eventually and inevitably delivering us freedom and progress. As Keedus points out:

> ... [Barth called] political modernity ... 'a snake forever biting its own tail' – whereby he referred to the *ouroboros* symbolism used on the original print of *Declaration des droits de l'homme* from 1789. In its attempt to overcome the absolute God, political modernity cemented the absolutism of the state, Barth argued, which in turn propelled it into the 'completely closed circle' ... of 'violence' from which 'we have no means of escape'. As Barth's readers know, an escape in fact existed – not to be found within the circle, but granted to us by the 'tangent' ....[133]

Keedus notes that Barth's "refusal to celebrate history" as "the workings of God through mankind" in typical Romantic fashion,[134] and his fierce objections to the romantic, Schleiermachian, theological Protestant treatment of history as moving toward progress and development,[135] went hand in hand with his attention to the significant harms to one's fellow humans that could be overlooked or excused by claiming history was a kind of logical, rationalized [logarithmic] system.[136]

Barth's rejection of the ouroboros is also suggestive of his recognition of what I will offer is the intensely "phagic" quality of the romantic and

[132] Cf. also the approach to nature discussed in Amy M. King, 2010, Quietism and Narrative Stillness, *Common Knowledge* 16: 532–551.

[133] Keedus, 'As A Tangent,' 55; Keedus is also citing Karl Barth, 1959, *Protestant Thought: From Rousseau to Ritschl,* New York: Harper, 30; and Karl Barth, 1968, *Epistle to the Romans,* 2nd ed., Oxford: Oxford University Press, 187, 156–7.

[134] Keedus, 'As A Tangent,' 55.

[135] Keedus, 'As A Tangent,' 39f.

[136] Keedus, 'As A Tangent,' 55–56.

CM "mere weaving of weaving[137]" (cf. Chap. 1) view of reality, demanding motion, change, proof-challenges, measurement, experiment, algorithmic progression, and so forth, to sustain the illusion that its view is really faithful to the existent world (while we in actuality, through all this motion, are constantly making it so ourselves, now in the postwar period, and especially after the Second World War, via ever more impactful techno-industrial and informatic interventions). The society which only has this weaving of weaving view must, in Jappe's formulation, eventually devour itself,[138] as it runs out of the physical and spatial means to sustain the illusion of having this weave really "be" and sustain existence. The "weaving of weaving" view of reality, as Jacobi characterizes it (cf. Chap. 1 above), is always in itself hollow of any means of upholding itself except to increase in speed and transformative capacity; eventually, it ends by reaching a point where it has no other way to hasten or expand its weaving, and so must consume itself in the effort to remain active and cogent as a picture of reality (more on this below). The only thing that opposes and limits the ouroboros is the tangent line that can break its self-contained logic at least at a single point; either the fuller part of reason, within the person, or the Other and really external, beyond us; either can reveal to the ouroborous its "fabricated" quality (Jacobi's term, cf. Chap. 1). The capacity to locate and encounter this Other in ourselves or in the world (or God, for Barth), and use this encounter as part of a basis for examining the ouroboros of our own necessitarian weaving, for holding it apart from our fuller judgment to see and understand it as a finite fabrication only useful in some limited kinds of cases but never as a real ontology of being, can be called, in opposition to the devouring and eating phage or phagic, the "phanic", or "seeing," capacity. Part of what Jacobi and Barth each, in their way, seem to offer as their stand against phagic naturalism – which expends our personhood in the wrong way just as it erects a false world – is the phanic person's. It is arguably the phanic capacity in our fuller faculty of reason that Jacobi and Barth are both most intent on trying to preserve, since they both understand that it is a precondition for understanding or acting upon their other arguments. There is potentially a significant environmental correlate to rebalancing the phagic with the phanic within the person, as we will continue to discuss.

[137] Jacobi, Letter to Fichte, 511 (no. 23).

[138] Anselm Jappe, 2023, *The Self-Devouring Society: Capitalism, narcissism, and self-destruction*, trans. Eric-John Russell, Brooklyn: Common Notions.

The outcome of the Second Controversy iteration was, historically, once more the defeat of the Jacobian-Barthian position in mainstream Western discourse. In the aftermath of the Second World War, religion and theology themselves faded further from this discourse, especially in terms of positions which Jacobi or Barth would find important. An ever-increasing insistence on the CM universe as the only nature we knew or needed came to dominate the Western understanding. Neo-Spinozistic "nature" was all that "was." The reliance on scientific versions of Spinozism became ever stronger; the slippage – sometimes intentional, sometimes accidental, sometimes tacit, but never fully explicated or even acknowledged as important – between the two types of necessity claim (cf. Chap. 2), increasingly with the third, artistic, type also present (but transmuted into more of a technological-production key – where X cannot be otherwise than as this machine makes or reveals it to be – than strictly speaking a solely artistic one), turned CM "nature" ever more concretized and totalized, even as it attracted new rounds of neo-Spinozistic theological and religious interpretation by pantheistic or panentheistic theologians now willing to shift their language and framing ever more fully into science's terms (we will return to this point below), just as philosophy had done a generation or two before.

At the same time, the West moved toward an inevitable third Spinoza Controversy "configuration", to again use Lazier's term. Anthropogenic environmental decline accelerated, as the West's effort to really inhabit CM's nihils of pure reason – to enforce its worlds of necessity as the only real worlds, using all the phagic force available to it as it successfully tunneled into the infinitesimal piece of real nature represented and mapped by laws of force and necessity relations, and harnessed the powers of force it obtained from nature as a result – brought on the widespread removal and conversion of the world as it was in itself and external to a now-burgeoning, exponentially exploding, human phagic devouring hunger. The West's decision to nonetheless do away with any last legacy hesitancy to shift its intellectual discourse solely into the key of necessity, do away with any lingering doubts as to whether the same type of knowledge which had generated horrors in two Continental wars and left humanity burdened with nuclear weapons was also the only form of knowledge that could put us in contact with the real, while also providing the basis for a new mythology of reason to animate society toward an ever more tight and total embrace of the phagic way of being on earth – in other words its continued committment, after the second Controversy and the Second World

War, to seeking to inhabit only the CM world, despite the removal of the real the West was accelerating in environmental terms – soon also led to another consequence which would further unsettle any appearance of a truly final resolution of the Controversy the second time around.

It turned out that the scientific ontology of logical, inherently identitarian, necessity was not the permanent and stable inheritor of the theological-religious understanding of our single world it sought to convey itself as during the first two Controversies. Instead, the CM scientific version of the "ontology" of necessity – the ontology which asserts about the world only "X cannot be otherwise" in the logical sense, while also smuggling in "X cannot *be*, otherwise" as it suits it, while still manipulating and testing both statements through mere theoretical or mathematical logic alone – revealed itself to be grotesquely, even potentially fatally, imitable. An explosion of pseudo-worlds drastically less robust, general, and nature-oriented, and drastically more specific and social and even individual than science's own, found ways of enforcing their own axioms and premises as the source of new and enforceable worlds – again, now at the societal and individual scale – "made" from necessity relations alone. What Jacobi might call internalism loosed on the world as absolutized naturalism, in other words, did not – and perhaps could never – stay confined to the practice of naturalism itself for long.

This proliferation of rational worlds, each with their own web of necessity relations which must be taken as literal and inexorable directives by the human subject living within them, has intensified the pressure on the twenty-first century West to confront its third Spinoza Controversy, to face its own either/or. The West must face the question one last time of whether humans can, or will, choose to turn away from CM absolutism back toward the truly external. A third and arguably inherently final crisis-point for the West to make one last decision on the deployment of necessitarian regimes, on the right use of necessity-based reason itself, has arrived: our third, inevitable, environmental Spinoza Controversy, one which indeed suggests it may have always been better named the Nihilism Controversy.

CHAPTER 4

The Spinoza Controversy's Twenty-First Century Return

In this chapter I will lay out two main aspects of the West's 21st-century critical monism (CM) which in our own return of the Controversy must be, as Sandkaulen said characterized Jacobi's original effort[1], contradicted and not refuted by the neo-Jacobian position. The first is the [arguably incoherent, when taken as an account of the entirety of nature] ontology of necessity ostensibly contained in big bang cosmology and underpinning the entire Western picture of nature as such, and the second is the accelerating human conversion and vanishing of the living (and even some of the nonliving) layers of the real external world. While I will intersperse some neo-Jacobian responses in this chapter, I will reserve most of my [still-nascent] neo-Jacobian position for the next (and last) one. These final two chapters are not intended to be exhaustive accounts of the twenty-first century return of the Controversy. Further refinement and exploration of why and how the Controversy has returned in the contemporary period in a specifically environmental key will hopefully unfold both in responses from others to this work, and in my own further future work on the neo-Jacobian position in what will hopefully form the topic of a sequel volume to this one.[2]

[1] Sandkaulen, Jacobi and Spinoza, 47.
[2] *Rural Philosophy*, forthcoming.

K. C. Snow, *Our Inevitable Third Spinoza Controversy*,
https://doi.org/10.1007/978-3-032-03045-0_4

Romantic Nature, Redux: Critical Monism's Big Bang Universe

In the early twenty-first century West, mainstream discourse (both in and beyond the academy) takes it as true that our real universe is somehow perfectly ontologized and fully existentially accounted for – at least in principle, even if details of certain theories may change – by scientific cosmology and the mathematical physics on which it relies. CM's necessitarian naturalism has arguably been building uninterruptedly since it vanquished Jacobi's position around 1800; it has even more clearly been established as the only serious view of nature beginning with – and in the century since – the time of the Controversy's second iteration, when another question mark raised as to the "fate of reason" (to use Beiser's formulation of the original Controversy), this time as Europe reckoned with the First World War in its aftermath, was in the main answered by a resurgent wave of new enthusiasm for Spinozist conceptions of nature and knowledge, with logical empiricism turning Western philosophy's face toward science just as Einstein and his colleagues began cementing mathematical physics' status as the necessary and sufficient underlying language of the new monism by launching scientific cosmology in its current form.

In the century since Einstein's 1917 paper, the CM "universe" has accreted many layers of interpretation and integration of new observational data, but its essence – the type of thing it is supposed to be, and the types of inputs scientists use in adding to or refining their image of this thing – is the same. The current, early 21st-century version of the scientific cosmological model of the universe claims to offer "a comprehensive description of nature, in which a set of primary constituents (radiation, matter, and an unspecified dark energy) are assumed with a partitioning determined from fits to the available data[3]". The model "posits that the universe began with a hot big bang, followed by an expansion calculable from the solution to two rather simple differential equations[4]". Though as of the early twenty-first century there is still no single precise version of the model which has been proven to be correct, strictly speaking, cosmologists generally converge on accepting a hot big bang, a place-holding Lambda to stand in for a "dark" (unobservable, only inferred) energy which drives the universe's perceived complete, all-encompassing

[3] F. Melia, 2013, The $R_h = ct$ universe without inflation, *A & A*, 553: 2.
[4] Melia, The universe, 1.

expansion, and cold "dark" (again, unobservable, and only inferred) matter; it is usually referred to as the Lambda cold dark matter (λCDM) model. The general outline of the resulting narrative is well known: around fourteen billion years ago, "all the matter, radiation, and dark energy[5]" that exists was unimaginably denser and more compact, first in a plasma-like state, and then undergoing a series of violent, high-energy phase transitions, soon generating the radiation which would become the Cosmic Microwave Background (CMB), the 2.7 degree Kelvin bath of "relic" light distributed apparently evenly throughout space. Because the CMB follows the black-body radiation curve, it is interpreted as showing us effectively that the observable universe in its entirety was once itself behaving like a single black-body,[6] sometime very close to the "t = 0" "beginning" of our observable universe (or, put another way, somewhere close to the earliest singularity back to which all of our observable universe's radiative energy (including its matter, "dark" matter, and "dark" energy) and gravitational force can be causally linked in the model).

The same massive energy transformations which ostensibly left the CMB as the oldest detectable layer in our sky's evolving palimpsest of observable radiation then are imagined to have continued to cycle through different states (all while the universe as a complete whole underwent cosmological, universal expansion or stretching), eventually allowing atoms, then hydrogen and helium, to form; from these then came all the other things we can see: galaxies full of stars and their planets, with the Milky Way, our sun, and our earth eventually among them.[7] As cosmologists Liddle and Loveday note, this (quite Schellingian) account has become the "standard view amongst cosmologists," who, again, assess

[5] Melia, The universe, 2.

[6] Cf. e.g. Liddle and Loveday: the "spectrum of radiation emitted by a black-body" – i.e., a "perfect absorber and emitter of radiation" – "has a universal form known as the Planck spectrum," Andrew Liddle and Jon Loveday, 2009, *The Oxford Companion to Cosmology*, Oxford: Oxford University Press, 29.

[7] The size of the observable universe in the model is the finite (envisioned as having a finite past historical span of existence, approximately 14 billion years) yet unbounded (because ceaselessly expanding and without an edge or an end) volume currently usually approximated as something in the order of 8.8×10^{26} meters. Cf., e.g., Paul Halpern & Nick Tomasello, 2016, Size of the Observable Universe, *Advances in Astrophysics*, 1. They find for a radius not a diameter, and give their calculations in Mpc, not meters, so I have converted accordingly.

The "finite" requirement envisioned by Einstein is also met by the fact that "the speed of light gives a maximum rate to the spread of information," cf. Liddle and Loveday, *Companion*, 314.

that "the Universe is described by the big bang cosmology, where the Universe has expanded from a hot dense initial state"; this model is "almost universally accepted by cosmologists as the best existing explanation of the observed Universe[8]". Cosmologist Aguirre notes that, however, "there is no sense in which the universe [as a whole, or as all that exists] is at any time 'small' or 'at a point' "; at "our" t = 0, a "whole infinite system" "springs into being all at once.[9]" (This is in keeping with the romantic and CM requirement that nature be a single whole, with nothing "outside" it, and in keeping with the way romantic views of nature utilize an ever-present, infinite and eternal, Nature when needed, including to smooth over paradoxes involved with claiming to know something infinite through conditions and states of finitude – as we will return to below.) Cosmologist Silk writes, "If we trace the evolution of the universe back through time, the universe becomes progressively denser and hotter. The … *observable universe* … becomes smaller and smaller … restricted to the distance that light can travel during the time elapsed since the big bang; the actual universe is much larger[10]". As is starting to become clear (and will become more so below), scientific cosmology speaks about the universe in two ways: as an infinite whole ("all that exists," or "actual universe", in the words of the cosmologists just cited), which is unobservable both beyond our part of it, and unobservable as such, if considered a singular entity or entire unity; and as our observable universe, which is finite for the purposes of science and observation, but also not in any way separate from, or other to, the infinite unobservable whole. It is finite in that the finite speed of light, combined with [current] interpretations of cosmological redshift as indicating that the universe is expanding (and expanding absolutely, as an entire entity, not only in some aspects), allows scientists to extrapolate back through a finite amount of past time and reconstruct what they assess must have happened to cause the expanding array of matter, energy, and gravitational force—and all the laws of gravity, atomic behavior, etc. they understand as being what "makes" this array – we observe through telescopes and on earth today. Because the singularity

Note, again, as we will discuss in what follows, that the universe beyond the observable envelope is usually (almost always) assumed to be infinite, though what exactly this means is variously interpreted.

[8] Liddle and Loveday, *Companion*, 314–315.

[9] Anthony Aguirre, 2011, Cosmological Intimations of Infinity, in *Infinity: New Research Frontiers*, eds. M. Heller and W. H. Woodin, Cambridge: Cambridge University Press, 177.

[10] Joe Silk, 2001, *The Big Bang*, New York: W. H. Freeman, 105.

with which the account begins its more realist segment of its history (there are also highly idealized extensions of this narrative back beyond $t = 0$, as we will see below) is effectively a romantic infinite chaos, still pre-formal and graspable by reason, but only barely (cf. Chap. 2), everything we see in our observable universe today must be accounted for essentially *ex novo* "beginning" (really, inferentially ending, and beginning from our present-day ontological theories) at this point. Thus the historical account of the past approximately 14 billion years also, in true romantic fashion, provides a cross-section or complete catalogue of what scientists regard as nature's core ontological makeup in the present and how such things are ostensibly made or arise beginning from a completely chaotic condition like a singularity (which itself is largely ideal and mathematical) or just "after" one. As we will continue to see, this complete indexing or cataloguing function means bang cosmology (and the physics upon which it is based) serves to frame and ground all the natural sciences; it is the cosmologists' "nature" to which all the other sciences must ultimately refer, and which in this way most represents the romantic ideal of a single unified science (at least in principle). And again, in true neo-romantic and monist fashion, there is no disjuncture between the infinite and the finite faces of the universe or nature within cosmology, as we will continue to see in what follows; they are two aspects of the same whole, a neo-Spinozist (not Spinozan, as we recall from previous chapters) Nature-nature.[11]

In order to generate the calculable universe models (originally centered on predicting how gravity acts, as we recall from Chap. 3, but quickly expanded to incorporate radiation phenomena like Hubble's observation of the redshift of distant stars' light which led to the theory of universal [Hubble] expansion, despite there still being no theory which joins the forces of gravity with those of electromagnetism) which ostensibly represent or symbolize our finite, observable universe in something like its entirety, the scientist must begin, as Jacobi would not be surprised to see, with the removal of anything which is not included in the theory. A "cosmological model … is normally a collection of physical laws assumed to be true, plus a set of assumptions as to the environment in which those laws

[11] In this work I am offering a compressed version of my argument that big bang cosmology has revived a neo-Spinozist and neo-German romantic "nature"; for a more in-depth version cf. Snow, Katherine C., 2021, *Transforming the finite: Toward a Jacobi-ian ecological critique of big bang cosmology's new German romantic philosophy of nature* [Doctoral Thesis], Edinburgh: The University of Edinburgh Research Archive (ERA).

are to be applied.[12]" What for Jacobi would be merely a sophisticated new version of an ideally-constructed "complex that we call nature[13]" is now ever more precise, a perfectly designated, fully testable, schema. As astronomer Chaisson puts it, essentially the scientist puts in values for variables and constants like mass, volume, pressure, etc., and "'Numerical experiments' ... crank out ... universe models[14]" based on Einstein's field equations and a few other mathematical calculations. One ends up with something like a bare model of how these values, which one inputs, change as an interrelated grouping over time. By toggling back and forth between the mathematical models and observational data based on what astronomers can observe in the sky, a rough life history of such a mathematically defined universe, which reveals (or proves) itself through its imagined evolution, is proposed – and, again, largely accepted. Most commonly, universe models also require simplifying the universe for the sake of calculation into "a single material which satisfies the assumptions of homogeneity and isotropy", which then accommodates using "the Friedmann equation governing the expansion of the Universe[15]". As Kroupa puts it, a hypothesis underlying big bang cosmology which is "so fundamental that it is usually not stated" is that "all present-day matter is created as a relativistic fluid during the hot Big Bang,[16]" and this fluid is initially modeled as a single, uniform volume.

A seeming requirement for this form of world-building, to make sure the universe being described or symbolized can meaningfully be said to include "the totality of all that exists[17]", is, as this process shows, to make it testable; this establishes the universe, in romantic fashion, as knowable and definable via change alone twice over, intensifying its malleability and intelligibility even beyond the manner in which the big bang conception already posits nature as having gone through an extraordinary, even absolute, degree of change over time. The assumption that "the totality of all

[12] Liddle and Loveday, *Companion*, 77.

[13] Jacobi, Concerning the Doctrine, 376 (no. 425).

[14] Eric Chaisson, 2006, *Epic of Evolution: Seven ages of the cosmos*, New York: Columbia University Press, 49.

[15] Liddle and Loveday, *Companion*, 77. Again, this model is the basic foundation of big bang cosmology to the current day, though some elaboration, including related to presumed quantum aspects of the initial singularity, has been added.

[16] Pavel Kroupa, 2012. The Dark Matter Crisis: Falsification of the Current Standard Model of Cosmology, *Publications of the Astronomical Society of Australia*, 29: 395.

[17] Liddle and Loveday, *Companion*, 314.

that exists" could ever be the same as the laws of physics and a schematic of some environment – an environment conceived as somehow resulting sheerly "from" these laws, or from the bare requirement that they be instantiated or be shown to have bearing – asserts natural law as inherently ontological, and realist, even as it is perfectly intelligible and ideal-mathematical, and, again, rests on change and test, as these are the ways in which laws make themselves known. The fact that we could never know all that must be the case (*other* than the bare fact of some laws conceivably bearing or holding true) in order for the world in which the laws work to really be present and exist, is entirely ignored. The [impossibly pure] imagined observation of a law holding (designating a certain subset of necessity-relations as potentially possible and potentially realized) and nothing else, means that the "presence" or "absence" of different laws (or parameters, or other calculations) takes on the connotation of the presence or absence of different worlds.

Nature remains aloof from the model itself, but also somehow directs and frames it in some unclear manner: without mechanism, but still seeming to anchor the statements of the scientist to the effect that some overarching Nature "would do" X or Y, via the presence or absence of law A or B, in order for a *universe like* J or K to result. Big bang cosmology does not offer a universe in a physical, Aristotelian sense, with nesting spheres and a cosmography; the spatial element of the big bang universe is always, on the large scale, stochastic, generic and placeless. There is no pervasive invisible curvature we have detected to date; we instead seem to inhabit a "Euclidean" kind of flat spacetime (which in turn bothers cosmologists as it does not seem as though "Nature *had* to" generate such a universe – more on this "flatness problem" below). Once the model seems to suggest infinity beyond what is observable, as is the case with this flatness itself, again an offstage Nature generative of our observable nature can cleave these perfect generics off onto itself, allowing our claims about our observable finite-seeming universe to remain both continuous with this infinitude and still rationally theorizable and manipulable.

All three features discussed in the first two chapters (and touched upon again in Chap. 3) distinguishing early German romantic ideas of Nature-nature can be discerned again here in big bang cosmology: (1) the distinctly romantic and neo-Spinozistic quadratic "nature" forged via combining realism and idealism, and ontological claims with the measure of our own epistemological capacities (or combining claims about the contours of being with claims to map the correct contours of legitimate human knowledge of nature), which we are trying to begin to represent in this section; (2) reliance on abstract speech of an eternal and infinite Nature seemingly beyond the universe model and grounding (or emanating) our seemingly finite observable universe, which we have also been presenting in this section and will continue to sketch in the first subsection to follow; and (3) the creation of a new (third) version of the early German romantic "new mythology of reason" drawing from knowledge of this new cosmic Nature-nature. The rough shape of this third romantic new mythology will also emerge in these earlier sections, but I will also briefly expand on its outlines in the second subsection to follow.

As we recall, early German romantic philosophy of nature returned again and again to the question of the human mind's relationship with infinite Nature. Conviction of our mind's access to just this type of Nature is arguably embedded in the CM position undergirding scientific cosmology as of the early 21st century. Cosmologists have embraced ontologizing the entirety of what is using statements of necessity which create logical and deductive maps or webs connecting together certain types of extractable regularities in nature, speaking methodologically, but which are interpreted as accounts of things actually coming to be, including for the first time (or the first time "after" the singularity). While perhaps even Jacobi might not have been able to foresee the extent to which big bang cosmology has carried out this approach, the essential gist of how the natural universe's grasp and reconstruction by reason is justified or explained remains quite close to the premises and steps for which Jacobi critiqued the Spinozists of his day. With the universe model so unestablished and changeable in its specifics (and, in multiverse theory, as we will see, endlessly proliferating), an internalist idea, "Nature," is clearly premised first (and often referred back to), serving as the imaginary ground on which the theorist must stand to undertake his experimental "re-construction" of ostensibly all of nature. Nature is not Other to the nature the theorist weaves; the type of Nature being assumed means the theorist is claiming to be in contact and concert with the one infinite whole; this totalized access to reality is what allows him to give his construct the august title of "universe." This Nature, this claimed CM version of the absolute, encompassing all, also papers over everything not actually accounted for in the positive parts of the theorized "universe" (a gap as large as all of reality, in Jacobi's view) and allows the naturalist to claim this giant remainder for reason just the same, under "Nature's" title or aegis, and in this sense, Nature reveals itself as a Nihil of reason's own making, just as the nature-world the theorist constructs of knowable necessity can only arise after the naturalist first annihilates the real one (cf. Chaps. 1 and 2). For Jacobi, Nature-nature is better understood as Nihil-annihilator, and moving into the certainties of physics (which for Jacobi only "reveal" a little of existence, and do not allow us to "make" it) does not change this. CM's Nature which serves so many uses for the cosmologists cannot be modeled but can still apparently secure, and play a role in, the models; while the "universe" is phenomenal, and the changeable quality of its models shows this (as does the pressing forward of the modeling beyond the single universe and into the multiverse, as we will see), Nature, sometimes called reality or ultimate reality, is the ultimate frame of reference for the cosmologist. It remains one with theory because it is a CM type of Nature-nature with one claimed "principle": necessity, the hinge between reason and world used in German romantic and interwar monism now returned once again to perform the same function.

It is this perceived ability to consider Nature[-nature], arguably, which pulls the thinking "I" of the cosmologist, per their own admission, beyond any one universe, beyond any one physical world, and into the fullest possible absolutizing of the claims of natural science. Carroll holds that "the fantastic success of our investigations into the physical world and the astonishing ability to predict events with exquisite precision speak to the fundamental intelligibility of the universe", and that while some may call this descriptive rather than explanatory, "that seems like a category error. … the world itself simply is. We are able to discover what it is, and, for more than that, it is illegitimate to ask[18]". This marks a clearer nod than is usually given toward the neo-German romantic framing of cosmology: he is effectively pointing out that he is not talking about being as such, and that he is operating by intuiting the existence of both nature and the human discoverer's mind as prior givens (with the two presumably also seamlessly unified) and then proceeding *in medias res.* Physicist D'Espagnat points out that while scientists usually do not bother themselves with the category of 'independent reality' *per se,* one cue to seeing when they do in fact mean the latter, independent reality, is when "they use

[18] Sean Carroll, 2018, Purpose, Freedom, and the Laws of Nature, in *Neuroexistentialism: Meaning, morals and purpose in the age of neuroscience,* eds. G. Caruso and O. Flanagan, Oxford: Oxford University Press, 307–308. Krauss also hints in this direction, cf. L. Krauss, 2012, *A Universe from Nothing: Why there is something rather than nothing*, Afterword by Richard Dawkins, New York: Free Press, 143–144.

the word 'nature' ... it can only mean 'independent reality[19]'". Cosmologist Weinberg dismisses concerns over different ways of defining "reductionism" because while these concern scientific or philosophical epistemologies as formalized, "I am talking about nature itself[20]".

The interest in, and pursuit of, uncovering ever more of the one *nature* (linked with and emanating from Nature), distinct from calculating models for the observable universe, is, again, part of what seemingly allows cosmologists to take their universe theories beyond the $t = 0$ point [roughly 14 billion years ago]. The CM scientist does not *need* an origin point in the theological or even typical historical-causal sense, since nature is actually always present (given the eternal and infinite Nature of the romantic, neo-Spinozist type producing it ceaselessly) and always at least implicitly premised as such; the larger effort becomes one of expanding and enhancing what Jacobi would characterize as the "complex that ... [they] call *nature*[21]" via increasing our knowledge of its approximately 14 billion years of cosmic evolution, predicting how the evolution will continue long into the future, and so forth. Cosmologists' expansion of their universe scenarios also go back beyond or "before" $t = 0$, as we will see, forming an even more idealistic arm of cosmological investigation that resembles a neo-romantic hunt for ways of showing more of Nature's own behavior as the self-grounding absolute, or as the nature- (or universe-) generator. Nature is also still ostensibly only an exhibitor of perfect necessity, and is also fully theorized through considering the imagined presence or absence of various natural laws themselves, as we will continue to see. The scientists' version of the romantic Spinozistic single, infinite, eternal unity is in this context sometimes presented as the highest target for all knowledge, now able to be subjected, as it were, to direct speculation (more on this below). Nature, again, is always the presumed-existent presence – something like a substance, but never defined (or limited) – and can be tested only as to whether certain laws inhere in it, and grasped only (somehow) as the seat of these laws. Krauss claims the multiverse (cf. below) is "like" a creator but more "logically consistent[22]". Some cosmologists, e.g. Davies, overtly name their generative, quasi-offstage, quasi-acknowledged, Nature as

[19] B. D'Espagnat, 1987, Empirical Reality, Empirical Causality, and the Measurement Problem, *Foundations of Physics*, 17: 527.

[20] Steven Weinberg, 1992, *Dreams of a final theory: The search for the fundamental laws of nature*, New York: Pantheon, 54.

[21] Jacobi, Concerning the Doctrine, 376 (no. 425).

[22] Krauss, *A Universe from Nothing*, 175.

God[23]. Nature is also often presenced in ways which compare it to a subject-like infinite mind, one that thinks and acts through laws to "make" the universe, even if it is never quite limited to this role nor, again, quite fully or consistently acknowledged or defined.

Generality and Infinity in the New Nature

Big bang cosmologists, in their efforts to constantly expand their representations of the "actions" of Nature "making" nature, are not impeded by the implication that this Nature is also presumed to be infinite both in space and time (and even to constitute a new version of the romantic absolute). Cosmologists resort to numerous neo-Spinozistic generalities or generics to attach "behaviors" or principles or "traits" to it, all without sacrificing necessity as its sole arch-principle. Faced with attributing necessity to Nature in order to use it as a convincing "maker" of the observable universe (somehow) simply through the bare presencing of laws which were in-principle observable, cosmology for many years relied upon a radical reduction of the historical-ideal $t = 0$ state of affairs to a type of metaphorical single idealized laboratory experiment. However, the single experiment notion was eventually deemed to make the instantiation of our entire observable universe *in germo* seem too irrational or uncaused – not "necessary" enough (cf. below).

The use of mathematical infinity was the first way finite descriptors and processes were ostensibly made able to describe a real[-ideal] physical infinitude, arguably simply via speaking in the perfectly generic or categorical key which mathematical infinity offers (again reminiscent of Novalis and Schlegel, and even Schelling). As we saw in Einstein's 1917 paper (cf. Chap. 3), the notion of conditions "at" infinity was used as a spur not to physically imagine infinity as a spatial reality or active place, but instead effectively to romantically "raise" finite categories to the level of inhering as characteristics of Nature (linking the finite with the infinite, as is necessary in monism), or in some way maximally test our finite constructs by echo-sounding them off of Nature itself, in order to assure ourselves that they inhered absolutely in things.[24] Our seemingly thinly disguised

[23] Cf. Paul Davies, 2004, Teleology without teleology: Purpose through emergent complexity, in *In Whom We Live and Move and Have our Being: Panentheistic reflections on God's presence in a scientific world*, eds. P. Clayton and A. Peacocke, Grand Rapids: William B. Eerdmans, 95–96.

[24] Cf. Footnote 61 in Chap. 3.

scientific versions of seeking a theistic mandate or guarantee for human theoretical speculation have, since around the 1800 watershed, usually involved nature in this way, drenching it with meaning ostensibly resulting from its display of perfect necessity, and imbuing it with an overseeing Nature, even as it was brought ever more completely within a rubric of scientific intelligibility and malleability. These were all devices which did not convince Jacobi any further of these constructs' reality, and even from a point of some neutrality we might begin to acknowledge that even our contemporary neo-Spinozist constructed universe seems at times to protest the meaning inherent in its necessitarian constructs a little too much, recalling Halmi's assessment that the German romantic movement burned brightly but then relatively quickly burned out, after having been *unable* "to secure any actual meaning from a purportedly infinite store of potential meaning" in the world as they conceived it (cf. Chap. 2). Mathematical infinity arguably does not truly "break in" (cf. Chap. 3) upon finite calculations or make them encounter (or apply to) something really infinite in the world; on the contrary, it effectively merely takes some finite calculative operation and makes it categorical or perfectly generic. It requires the CM framework to interpret the resulting generic as something like an absolutized or romanticized concept, capable of inhering in something like Nature itself. Yet just this faith (or belief) in the generic case as viably useful for attributing or attaching various principles or "traits" to infinite Nature (feeding into a neo-romantic claim along the way about the intertwined relationship between the finite subject or mind and infinite, unbounded or perfectly absolute Nature itself, just as Nassar points out was important to the German thinkers[25]) has been built into CM scientific cosmology since Einstein, as noted earlier, and as early detractors like Dingle tried to critique.

As we have already noted, "Nature" serves to tether the CM "universe" to a non-numerically interfering, yet [ideally-really] productive, ground, functioning not least to allow the scientist a cognitive standpoint from which to model the whole of the observable universe as though from without. The cosmologists then also blend speech of this idealized Nature with predictions about what it would "do" that would also lead to observable results, using certain natural laws' sheer presence or absence, again, as a stand-in for the act of "making" our observable universe.

As cosmology developed through the twentieth century, the neo-romantic question of how to adequately genericize necessity itself, how to

[25] Cf. Nassar, *Absolute*.

adequately infinitize it so it could inhere in Nature, such that Nature itself would have "had to" make [at least] one real universe precisely like ours *in terms of the natural laws (and thus the webs of necessity-relations) ours displays*, was gradually taken more and more seriously as a scientific problem. Cosmologists leaned toward seeking some type(s) of meta-necessity rule(s) bearing upon Nature as it made universes, just as necessity was the only thing constituting our universe itself once it had been made.

Still resorting to generics, and in pursuit of this meta-necessity they could attribute to Nature "naturing" in a neo-Spinozistic sense by making universes, plural (as the "activity" was now increasingly imagined as a class or type, not a single occurrence), cosmologists began discussing the notion of an infinitized generic meant to inhere in Nature called the "cosmological principle" or the ostensibly "Copernican" principle (CP), which states that the universe must be homogenous and isotropic on the large scale: it must lack a privileged directionality or orientation, presenting, at its total scale, perfect stochasticity, and also removing any possible hint of boundaries or edges to the universe, which, in true CM mode, is all that is (once we "include" the unobservable whole), so that it is impossible to refer to anything other than or "beyond" it. Committing to the CP sets the subject onto a particular path vis-à-vis the universe. Rendering the entirety of the universe something which must really first be found to serve out or manifest a rational, ideal "principle" (anathema to Jacobi, per above, cf. Chaps. 1 and 2) means this "universe's" intelligibility and ideality is forever after primary, placed over and above its reality. Romantic necessity's inclusion of aesthetic "necessity," intuited by the knower, among its three legitimate types of necessity-relation which can reliably show us reality (cf. Chap. 2), offers the romantic cosmologist an assuaging "best of both worlds" gloss on assuming the CP as true: nature, in line with its innate harmony and oneness with our minds, must show itself as amenable to theory, and thus [only] for the neo-romantic, nature does not suffer any distortion as a result of our beginning with claiming the CP is really true of the infinite universe. It seems to the scientist, aesthetically, intuitively, that nature "must" admit us to understand it as a unity, and via necessity-relations as theory reconstructs them. The CP, outside of the CM framing, is disingenuous to call a "principle" of Nature, since it arises due to an epistemological limit. It is a fact that mathematically, scientists *require* assuming the CP if they are to simplify, and thus be able to solve, the Einstein field equations which form a core part of the basic universe model.[26] Only in a CM view is there no contradiction between following the epistemological requirements and limits of the human knower, and claiming that where they lead us is to absolutely objective and totalizable ontological truths established and shown to us by nature. Taking the CP as given, and then also holding that the equations and models thus enabled also "show" us the "Copernican-ness" or stochastic uniformness (and unity) of the universe in reality, is only possible in the CM approach. The universe is embraced as both a simplified schema for the field equations, and real; it is seen as ultimately infinite, eternal, and boundary-less as Nature; finite yet unbounded, in Einstein's terms, and thus still connected with or derived from the infinite, as nature. The cosmologist gains a proximity to finally being able to occupy Nature's own standpoint of production [of not just one, but many, universes], the dream of the romantics. Cosmologist Davies stated in 2018 that "Linde's—eternal inflation, for example – … is a natural mechanism for making universes. One can dream up a universe-generating mechanism. That in itself is not a problem[27]". The historical Copernicus (among others) did change astronomical understanding by de-centering the Earth, cosmographically, of course, but he is also perhaps the beginning of modern science's claims to access an infinite *yet* intelligible Nature, which, if we recall Jacobi's concerns, must always be the most egregious type of internalist *nihil* – making any cosmology which claims to invoke it in fact the very opposite of "Copernican" in the commonly invoked metaphorical sense of seeing more objectively or de-centering the human view.[28]

[26] Per above, to allow these calculations to succeed, the scientist must assume that the observable universe as a whole is homogenous and isotropic (the claims of the CP), simplifying the universe into something like a (four-dimensional) "spherical" volume of perfectly smooth fluid for modeling purposes.

[27] Paul Davies, 2018, in Steve Paulson, Paul Davies, Ard Louis, & Lucianne Walkowicz, A touch of awe: Crafting meaning from the wonder of the cosmos, *Annals of the New York Academy of Sciences*, 1432: 50.

[28] Copernicus himself seems to understand the actual centering entailed in the "Copernican" move. Historian of science Hallyn notes that Copernicus believed in a kind of 'anagogy' wherein man could ascend to God's place and capture divine wisdom. Hallyn writes that Copernicus:

> postulated that this world was created *for us*, and *for our use:* "mundus *propter nos* ab optime et regularissimo opifice conditus," he declares … As the beneficiary for whom the world was made, man can attain *true* knowledge. In place of a universe whose beauty and rationality escape us, and which thereby calls us to humility, Copernicus substitutes a cosmos for which man is the final purpose and whose true plan he can reconstruct ….

After the discovery of the Cosmic Microwave Background (CMB) in the mid-1960s further bolstered big bang theory against potential detractors, and gave cosmologists another major observational data-set to work with, cosmologists continued to seek more ways of trying to use an off-stage Nature or meta-necessity principle to explain the very existence of our nature with its particular laws and not others. The observable universe as a whole entity was increasingly presented as not necessary enough. The goal became to hem in universe-making as an "activity" (again, carried out, implicitly, by infinite Nature, with the cosmologist the reliable reconstructor of this act), so that as a genre of action it was strictly governed by necessity and law. The simplicity imagined, in typical romantic fashion, as characterizing the early universe came to bear more and more weight in the account as a result; this type of simplicity was needed to begin to try to work universe-births themselves into a predictable, generic, fully necessity-defined (or lawlike) category of phenomenon. "Travel back in time," cosmologist Smoot writes, in a quite typical presentation, "and we see an ever-greater simplicity and symmetry, with the fusion of the fundamental forces of nature and the transformation of particles to ever-more fundamental components[29]". All the complexity around us today "flowed from a deep simplicity as matter metamorphosed through a series of phase transitions[30]". Weinberg, in his famous popular work *The First Three Minutes*, says of the earliest universe that it is "simpler and easier to describe than it ever will be again. It is filled with an undifferentiated soup of matter and radiation[31]". Weinberg writes that "during the whole of the first second [after $t = 0$] the universe was presumably in a state of thermal equilibrium, in which the numbers and distributions of all particles … were determined by the laws of statistical mechanics, not by the details of their prior history[32]".

However, after the heavy generics generated by mathematically infinitizing various laws, and the exploration of the CP, the appetite for more meta-necessity in the model continued to increase. Roughly speaking, discussions spread to why and how to include the anthropic principle (AP) in

Cf. Fernand Hallyn, 1990, *The Poetic Structure of the World: Copernicus and Kepler*, New York: Zone Books, 55.

[29] George Smoot and Keay Davidson, 1993, *Wrinkles in Time*, New York: Avon Books, 290ff.

[30] Smoot and Davidson, *Wrinkles*, 290ff.

[31] Steven Weinberg, 1993, *The First Three Minutes: A modern view of the origin of the universe*, New York: Basic Books, 102.

[32] Weinberg, *First Three*, 146.

models, and then began to consider the broader problem of fine-tuning; then support for the adoption of a theory of inflation grew, followed by increasing belief in some type of multiverse, which soon became an infinite multiverse.

The AP and fine-tuning concerns trace back to early cosmologist P. A. M. Dirac and his (and cosmologist Eddington's) type of quasi-rationalistic and idealist, quasi-realist ruminations in and around the 1940s on various numeric constants (or ratios) appearing in the universe models and how they might or might not evolve with time.[33] Another cosmologist, R. H. Dicke, revitalized these issues in 1961, asking whether one of the ratios with which Dirac had been concerned might be affected and shaped by the presence of human observers: in his words, whether perhaps "*T* is not a 'random choice' from a wide range of possible choices, but is limited by the criteria for the existence of physicists[34]". Scientists then in response put forward a principle, the AP; the AP can be equated with a claim either that initial conditions at (or just after) our causally connected universe began had to allow for natural laws to emerge such that these laws allowed for the emergence of observers (this is the "weak" AP); that "the only types of Universe permitted to exist are those which do contain life during at least part of their existence[35]" (this is the "strong" AP); or that observers actually *make* the universe directly, through quantum effects (the "participatory" AP, advanced by John Wheeler).

Though various of these three APs gradually became common components of scientists' methods of weighing different possible big bang models, they did not solve the perceived necessity crisis in big bang cosmology. The universe's "overall" Euclidean geometry which seemed to imply a "flat" large-scale structure seemed vanishingly unlikely given that scientists believed it could have had so many different degrees of curvature; and there was unexpected homogeneity in the CMB across too much distance. Despite the fact that the homogeneity of the overall structure of the universe had to be assumed to make the models possible, and was enshrined abstractly as the CP, cosmologists reasoned that the CMB itself and our universe's curvature, as both were believed to be products or traces of our

[33] Cf., e.g., P. A. M. Dirac, 1938, A new basis for cosmology, *Proceedings of the Royal Society of London, Series A,* 165: 199–208.

[34] A. H. Dicke, 1961, Dirac's cosmology and Mach's principle, Letter to the Editor, *Nature*, 192: 440.

[35] Liddle and Loveday, *Companion,* 17.

big bang, "could" have been and "should" have been more anisotropic (in the CMB case) or actually curved (in the curvature measurement case), not isotropic and flat, respectively.

Adding inflation, the speculative theory that hyper-fast vacuum pressure shaped our universe's first instants after $t = 0$, to the early universe model was the first step proposed to solve these (and other) appearances, in the cosmologists' eyes, that the initial conditions of the universe, instead of being determined entirely by knowable relations of necessity linked with regularities we knew how to extract in our own earthbound physics, had been "fine-tuned" or arbitrarily set, implying that the natural laws which according to the cosmologists really constituted and had actually made the universe were, in terms of their particular values or characteristics, merely accidental rather than [meta-]necessary, a combined prospect altogether anathema to CM (and its notions of Nature-nature). As one cosmologist put it, "we will not be satisfied until every physical measurement can be predicted from theory alone[36]". Inflation was proposed by cosmologists like Linde and Guth in the 1970s and 1980s as part of a way to explain how our universe happened to hit the just-right mark at the time of the big bang that led it to have critical density[37] and thus a flat geometry, CMB isotropy, and its observed scarcity of magnetic monopoles,[38] again, mostly in the aftermath of observations of the CMB.[39] The theory of inflation holds, in cosmologist Greene's words, that "in the

[36] Luke Barnes, 2017, Testing the Multiverse: Bayes, fine-tuning and typicality, in *The Philosophy of Cosmology*, eds. Khalil Chamcham, et al., Cambridge: Cambridge University Press, 457.

[37] Critical density is the "particular value for the density of the universe which has the special property that it corresponds to a flat spatial geometry", cf. Liddle and Loveday, *Companion*, 82.

[38] The prediction that magnetic monopoles would dominate the universe per the standard model of particle physics, combined with the observation that they do not, is another situation the cosmologists considered unlikely or fine-tuned such that some mechanism must have intervened they had not thought of yet. Guth's inflation addressed this perceived problem too. Inflation was a "burst of rapid expansion, taking place after the monopoles formed," leaving them scarce in the overall makeup (so imagined) of the universe, and thus solving the monopole "problem" as well as the "problems" of the CMB's isotropy beyond the causal connectivity horizon and the flat spatial geometry of the universe per observations. Note that what is called a "problem" is usually an indication that some observed trait of the universe (ostensibly qua single whole) does not appear strictly necessary and instead seems merely arbitrary, unlikely, intelligently designed, or causes some other aesthetic or intuitive red flag for the cosmologist. Cf. Liddle and Loveday, *Companion*, 198.

[39] With its apparent coordination of homogeneity across "causally disconnected" regions.

universe's earliest moments, space would have swelled so fast that regions would have been propelled apart at greater than light speed[40]". As Krauss describes it, "the universe could have expanded during this inflationary period by a factor of more than 10^{28}. While this is an incredible amount, it amazingly could have happened in a fraction of a second[41]". Inflation was developed for its potential, as cosmologist Barnes states, to "provide a deeper explanation for some aspects of … [the standard model of cosmology] that seem unnatural or fine tuned.[42]"

Inflation too, however, soon led in turn to concern that this phenomenon itself had to be posited as unceasing and omnipresent in order to make it feel sufficiently necessity-bound, or as historically necessary, in effect, as everything done by and in Nature must be. The issues with positing a single incidence of inflation were part of the whole genre of problems cosmologists were finding in their model. Cosmologists were still uncomfortable with how "the standard models of particle physics and cosmology are both rife with numerical parameters that must have values fixed by hand to explain the observed world", because unlike the physicist, their job (as they saw it) was to explain these constants themselves, and their related laws, not merely to identify and apply them; the fact that "the world would be a radically different place if some of these constants took a different value[43]" was increasingly interpreted as a challenge to the cosmological claim to explain all of nature and what and how it was. Arguably the fierce realism intertwined with the mathematical idealism of these models contributed to the scientists' flat refusal (or complete inability due to the terms of the models themselves) to consider as an alternate possibility that [radical] human epistemological limits prevented them from even approaching describing (much less needing to worry over how to predictively pinpoint within a canvas of infinite other really possible but not realized outcomes) the raw ontological "generation", primal making, or initial "setting" of our laws and parameters. Such self-sustaining, world-generating power being assigned to laws of nature, the elevation of them to such an important ontological, cosmological role that they required an

[40] Brian Greene, 2011, *The Hidden Reality: Parallel universes and the deep laws of the cosmos,* New York: Vintage, 51.

[41] Krauss, *Universe From Nothing,* 97.

[42] Barnes, Testing the Multiverse, 447.

[43] Anthony Aguirre, 2007, Making predictions in a multiverse: conundrums, dangers, coincidences, in *Universe or Multiverse?,* ed. Bernard Carr, Cambridge, UK: Cambridge University Press, 367.

additional pedigree of meta-necessity or meta-ontology in turn, was unique to the CM world construct. For a thinker not pre-committed to the CM quadratic version of "nature" – malleable, intelligible, accessible via intuition, and always backed, grounded, and non-mechanistically directed by an infinite pre-existent which was also sheer necessity – it might seem quite unremarkable that the models lacked meta-necessity. Since all cosmology's principles or theorized phenomena or traits held to be true of the imagined universe "as a whole" (whether on its finite, visible side and/or on its infinite, ideal side) are really just statements of some categorical extrapolation from some physical law (including by "taking it to infinity" in Einstein's sense, i.e., genericizing it) or finite set of observations, there appears to be no way to make a permanently stable or complete model that meets the CM bar; as soon as one generic is stated, it can be perceived as being selectively fine-tuned instead of arising out of pure [meta-]necessity. Cosmologist Rees identifies six parameters which "constitute a 'recipe' for a universe[44]", writing that "if any one of them were to be 'untuned', there would be no stars and no life[45]". Nature, in this metaphor, it seems, is the [offstage] chef; note here the romantic CM claim that our knowledge extends beyond the cake, beyond the recipe, and into the mind of the chef and his discerning of not only which ingredients were really needed to bake a universe, but also which ingredients could be varied, and in which ways and ratios, in order to potentially make different flavors of universe other than our own – none of which we would ever even in principle be able to sample, but all of which would still be commensurable enough with our own for us to be able to intellectually imagine tasting. Outside a neo-romantic framework, again, these preoccupations with maximizing necessity and discarding anything tuned seem nearly nonsensical, since surely anything at all which is defined solely by using parameters and laws, will be highly sensitive to imagined changes in or to these parameters and laws as one of its unavoidable characteristics. But the neo-romantic scientist does not perceive it this way. The universe is not the full story, the full story is Nature, and we cannot be correct if the image we are receiving of It makes it appear as though It "were" anything other than that which acts sheerly via necessity relations, and makes our universe via a natural history manifesting only necessity at each and every

[44] Martin Rees, 1999, *Just Six Numbers: The deep forces that shape the universe,* London: Phoenix, 4.

[45] Rees, *Just Six*, 4.

juncture and joint. The CM view also means, as noted above, that we need the account to show us, in true romantic fashion, how our universe arose not from ordinary chains of cause and effect or from states close (or identical) to its state today, but absolutely, from out of a sheerly infinite [yet romantically still-graspable] chaos of only pre-formal character, up to the present day full of quantum physicists and other complex life forms. Rees found six parameters not yet regularized or necessary enough; other scientists would find other too-arbitrary links in the imagined chain. Inflation, again, was a case in point of the root problem. The confusion it caused for the model (cf. below) arose from the structural problems with the scientists' way of trying to assert a generic (and infinite) case as a source of a specific (finite, historical) universe. This challenge was one of the German romantics' (and later C. S. Peirce's) preoccupations, we can recall, and in a different (negative) sense, since Jacobi considered the notion of generics really acting in the actually existing world, let alone "creating" it, to be nonsensical, also one of Jacobi's. Generic cases, as we have seen, are what allow scientific CM cosmology to make claims about the infinite and what it must be "like", and make universal-scale calculations reach predictive outcomes of a zero or a one value; it cannot do without them. Yet they unavoidably lead to problems when used as the "source" of our finite observable universe. Inflation as the cosmologists understood it was just a generic possibility in every sense, as it has never been observed.[46] Cosmologists merely assert it is one of Nature's capacities (and nature's phenomena) because it can be modeled using existing laws of physics. Once it was inserted into the model, however, its very auto-poietic character that made it so helpful to the model (and so helpfully generic rather than tuned – at least this was the thought at first) also meant it was only intuitive to posit it as a total and self-sustaining process: hence the model had to move to hypothesize that our universe grew out of "runaway" or "stochastic" inflation. Since everything in the monist naturalist ontology of necessity must be perfectly necessary, and act in keeping with the absolute and all its principles, runaway inflation (or eternal inflation, cf. Davies

[46] As Barnes puts it, "we do not know what the inflaton is", cf. Barnes, Testing the Multiverse, 448. Belyaev notes that the possibility of a vacuum "is generally accepted [to exist] but there are no … assumptions that can help us understand it. It is completely neutral, as if it is not there at all," and its use in first-moments cosmology represents only "a materialized mathematical environment", cf. Alexey Belyaev, 2017, A new model of the birth of the universe, *Universal Journal of Physics and Application,* 11: 186. (Since inflation happens faster than *c*, it is also presumably unobservable.)

quotation above) itself then had to be better infinitized and elevated, made as necessary and Nature-reflecting as possible; this led on to multiverse theory.

In an early foretaste of multiverse theory, cosmologist Tryon in a well-known 1973 letter to *Nature* says he seeks via physics the [perfect] 'predictive' recipe for the entire universe, and puzzles over the fact that "there is no apparent reason for such an event to occur[47]" where 'such an event' is, in his terms, the creation of the universe. Tryon wants to know *why* the Universe has "its particular values for energy, electric charge, baryon and lepton number and so on[48]". He reasons that "If it is true that our Universe has a zero net value for all conserved quantities" "then it may simply be a fluctuation of the vacuum, the vacuum of some larger space in which our Universe is imbedded," and if we are to "wonder how a vacuum fluctuation could occur on such a grand scale," his two reasons are that "the laws of physics place no limit on the scale of vacuum fluctuations", and his second criterion is that of the anthropic principle.[49] The "some larger space in which our Universe is imbedded" positions our universe as, again and here even more concretely, emergent from something primordial suggestive, unsurprisingly, of Nature. Smolin later calls this undefined larger space the "landscape[50]" in a generic, mathematical-probabilistic sense; Tegmark calls it an "infinite 'ergodic' space[51]" or elsewhere tends to simply presume as an ever-present and infinite expanse of space and time.[52] (Jacobi would surely point to it as our own theoretical faculty running out of descriptive power as it sought to describe an obvious mere *nihil* of reason to which it still sought to attribute meaningful properties, including perhaps tacitly representing it as Nature Itself.)

Inflation's pervasiveness and genericity eventually led cosmologists on to thinking of an infinite multiverse, where inflation would become the fluctuation that continually birthed universes out of this primordial

[47] Edward P. Tryon, 1973, Is the Universe a Vacuum Fluctuation?, Letter, *Nature*, 246: 396.

[48] Tryon, Vacuum Fluctuation, 396.

[49] Tryon, Vacuum Fluctuation, 397.

[50] Cf. Lee Smolin and Roberto Mangabeira Unger, 2015, *The Singular Universe and the Reality of Time: A Proposal in Natural Philosophy*, Cambridge: Cambridge University Press, 454f.

[51] Max Tegmark, 2007, The multiverse hierarchy, in *Universe or Multiverse?*, 99.

[52] E.g. in Max Tegmark, 2015, *Our Mathematical Universe*, Penguin Random House UK, 134, Fig. 6.4.

Nature-like layer, variously configured as a spacetime, a landscape, a vacuum, or other entity (sometimes even sheerly ideally conceived of as a pure mathematical probability-space). Cosmologist Greene says the "vastly new picture of reality's expanse ... can be grasped most easily with a simple visual aid" which is to "Think of the universe as a gigantic block of Swiss cheese"; he also includes a generic image of bubbles meant to illustrate this stratum, "an ever-expanding spatial environment permeated by a high-valued inflation field;" on the next page it's "an eternally expanding spatial expanse[53]". Krauss writes that "we don't currently have a fundamental theory that explains the detailed character of the landscape of a multiverse[54]". There is not too much agreement on what, or even "if", the multiverse really "is". The multiverse is only a generic; a [merely asserted, would-be] "principle" [of Nature] simultaneously claimed to be real – at least in the minds of some cosmologists; others are content to keep it as a schematic, in limbo between a new version of the principle of sheer necessity and a really real, infinite, physical and historical fact. Again, for Jacobi, the infinite multiverse would just be a dubious logical exercise in thinking about where the chain of requiring ever more articulations of necessity or perfect genericity (genericity which, as noted above, in its totalizing, infinitizing, or automatic character seems to somehow satisfy the notion of being fully necessary or un-"tuned" for cosmologists, and also to be the case or mode of description or positing most useful for ascribing (always necessity-based) "traits", principles, or properties to infinite Nature) would end if we could extend it infinitely. Tegmark identifies four simultaneous layers of multiverse,[55] Ellis four different possible kinds of multiverse which are not the same as Tegmark's,[56] and Carroll details the Everett-ian type.[57]

No matter what type of infinite multiverse is espoused, its purpose is to make our entire universe "typical", i.e., categorized and thus able to be explained (in the sense of being theoretically predicted or predictable) via a rule, or even directly made "of" and "by" merely the unrolling of some type of algorithmic necessity relation-set or law. Cosmologists' aesthetic

[53] Greene, *Hidden Reality,* 65–67.

[54] Krauss, *Universe From Nothing*, 176.

[55] Cf. Tegmark, *Our Mathematical.*

[56] G. F. R. Ellis, 2017, The domain of cosmology and the testing of cosmological theories, in *The Philosophy of Cosmology*, 26ff.

[57] Cf. Sean Carroll, 2019, *Something Deeply Hidden: Quantum worlds and the emergence of spacetime,* New York: Dutton.

self-interpretation of how and why they must go through such steps (or, for a Jacobian, such contortions) in seeking the origin of our universe is always that they must embrace the "Copernican" view; those who espouse an infinite multiverse of some kind often try to insist on its truth by pointing to how it "Copernican-izes" our view of the universe (and by implication moves us closer to full and complete access to the standpoint of production of Nature, the implicit maker of universes). Rees lauds the multiverse as heralding a "fourth and grandest Copernican revolution[58]" which makes not only of the Earth, or the human, or the galaxy, something 'typical', but also the *entirety* of the universe. As Greene puts it, "multiverse-based explanations" of the universe stand poised to lead us through "the natural step toward completing the Copernican revolution, five hundred years in the making[59]". Cosmologist Hartle presents a radical Copernican claim when he holds that perhaps the final generic necessity we need to articulate is one where the scientist essentially vanishes through relinquishing all temporal or spatial specificity in his observations or characterizations of the universe; he wonders whether the scientific thinker is potentially capable of relinquishing his reliance on *any* "features of our current theoretical framework" that "reflect our special position in the universe"; for Hartle these are only "excess baggage", rather than "fundamental", and should indeed come to be discarded:[60]

> in the future this might be seen as the time when scientists began to take seriously the idea that it was important to consider the universe as a whole and science as a unity, the time when they began to take seriously the search for a law of how the universe started, began to work out its implications for science generally, and began to discard the remainder of our excess baggage.[61]

Just as the universe is a finite manifestation of something that is best understood as a new version of infinite, Spinozistic Nature, the I's viewpoint will finally be overcome with all its inopportune specificity and its clash with the ideal of true, fully unified scientific understanding. Schlegel's

[58] Martin Rees, 2018, *On the Future: Prospects for humanity,* Princeton: Princeton University Press, 184.

[59] Greene, *Hidden Reality,* 357.

[60] James B. Hartle, 1991, Excess Baggage, in *Elementary Particles and the Universe: Essays in honor of Murray Gell-Mann,* ed. John Schwarz, Cambridge: Cambridge University Press, 2.

[61] Hartle, Excess Baggage, 8–9.

dream of freeing us from the thing-ly altogether (cf. Chap. 2) here finally seems to encompass even our own selves; such a pitch of romantic sublimation removes even Hölderlin's more embodied requirement that we take the aorgic into ourselves, and put ourselves, back into it in turn, on the path to knowing (or proving) ourselves to be one with the infinite (cf. Chap. 2).

Eventually, all these attempts to achieve knowledge as seen from Nature's own standpoint of producing nature or universes, and achieve coherent descriptors of infinity via chaining together ever more layers of perfect[ly necessary] generics interpreted via recurring re-statements of belief in necessity as the only ontological force of being and making in the universe, ran into an even larger problem.

CM cosmology may avoid the classic causal regress problem afflicting other necessitarian cosmologies due to its monism, its *in medias res* direct thinking of a quadratic, romantic-type universal nature, and its [tacit] reliance on an offstage and always existent Nature. But from Jacobi's perspective, we can recall, it is not the content of a rationalized ontology of necessity which is its first problem, but its essential form, and the antechamber through which it forces the subject to pass before it, as it were, begins to world-build using reason alone, the antechamber in which the subject must enlist its theoretical world-building capacity in a campaign (to rationally ontologize what really is outside the "I") it can by definition never win, not because of any flaw within the various types of rationalized webs of necessity per se, but because they are not the same kinds of thing as the real world around us, and only position us to meet, or dwell within, our own reason['s products]. And in big bang cosmology, eventually the ouroboros (cf. Chap. 3) meets its tail and must take the bitter bite. In the infinite multiverse, at last reason confronts the inevitable end-point of its "world"-building: the undermining of the very principle used as its only building material, as the only thread in its "weaving of weaving" construction – necessity itself. Necessity itself, qua would-be substantive principle inhering in Nature, gets taken up into its own "machinery" it loosed on the world in its own image, "necessity" not as a principle but as a process, a set of operations, a tool to render anything into a perfectly necessary case of itself, "raising" or "lowering" (in Novalis's words, cf. Chap. 2) it until it becomes perfectly categorical, perfectly necessary, and perfectly infinite. Necessity as a principle of Nature may retain some vestige of the grandeur Spinoza installed in his one substance; necessity as a rendering operation is its senseless opposite, and can operate to elevate anything at all to a necessary or absolutized case and thus the ostensible backbone of a "world" or ontology of necessity. If someone wanted to construct a necessitarian regime based on the color yellow, applying the operational version of necessity we can simply outline this world's "programming" or "law" as, to borrow from Voltaire's formulation from *Candide* (cf. Chap. 5), yellow is necessarily yellow, and all is for the yellowest in the yellowest of all possible worlds. The only case where the outcome does not work, as it turns out, is where necessity as an operation meets necessity as a principle. When scientific cosmology tries to necessitize necessity, in the form of the infinite multiverse, it ends up unraveling necessity instead, exposing its arbitrariness as both a "world"-building principle *and* as a world-defining tool, and by extension suggesting the entire CM "universe" is nothing more than an arbitrary internalist fiction, a story about necessity which can find no grounding and must end by collapsing under the pressure it inevitably puts upon its own construct.

The infinite multiverse can be understood as cosmologists' final effort to try to save the validity of the "ontology" of necessity. But it turns out to not instantiate only the imagined law-governed, perfectly necessary reality scientists were trying to preserve and to feature as the multiverse's most important product. The infinite multiverse as a type of process or operation is effectively a generic imagined embodiment of the operational kind of necessity, a perfect genericizing and "necessitizing" device (or even "program") which also itself exhibits perfect necessity and genericity. However when it is affixed to the universe as a whole which is itself primarily known as (or via) relations of necessity, a new image results in which necessity of both types, again, lose their meaning. The infinite multiverse as a place becomes a space wherein necessity itself, doubly absolutized and genericized, has become meaningless and lost whatever explanatory power it once had, collapsing altogether and arguably bringing down the CM cosmological system, all of which is devoted to ontologizing based on the imagined strength of necessity as the primary tendon of reality, along with it. Tegmark's view is that there is a totalized infinity which crops up in even the very first of his four levels of multiverse, writing that the apparent Euclidean flat infinity (stretching beyond our cosmological observational horizon) of the universe, on its own already means we can infer that there are:

> infinitely many other inhabited planets, including … infinitely many copies of you – with the same appearance, name and memories. Indeed, there are

infinitely many other regions the size of our observable universe, where every possible cosmic history is played out.[62]

We can already see here how necessity itself, subjected to necessitization, begins to defeat the necessity-constituted Nature-nature the scientists claim to be uncovering. Guth conveys the problem succintly in a 2014 technical article: "In an infinite multiverse, we do not know how to define probabilities … anything that can happen will happen an infinite number of times[63]". This is arguably a definition not of necessity but of its complete absence, leaving the scientist, with his neo-Spinozist version of Spinoza's posited a priori claim (according to Schelling, cf. Chap. 2 above) of the world as the set of all possibles, now confronted with an infinite repetition of these possibles which removes, arguably, their traditional status from them, and makes of the universe an absolute chaos: not the romantic, pre-formal, as in pregnant with possible form, chaos, but a post-formal chaos, one from which we cannot retrieve any sense even in principle.

The infinite multiverse thus reveals the always-present limits on ontologizing the universe as a whole "as" or as "made via" the operations of necessity alone; necessity does not allow us to leave the closed loop of our own theoretical identitarian reason, and indeed only shows us its ligaments, not the world's – ligaments which inevitably shift us perilously ever-closer to senselessness and unreason if we forget their circular bent (as Jacobi so presciently tried to warn). Using logically-derived necessity relations to try to ontologize a real world and the very conditions of its complete existence or presence can only ever mean deducing from some chosen axiomatic premise or entity X, and some abstract "necessitizing" rule Y, a resulting "world" Z, which can never be *real.* Z will always be only an idea, a dream of reason; an imagined notion (what Jacobi might call a "fictitious concept", per Chap. 5 below) that a really existing Nature somehow creates the actual world "through" the unfolding of the kind of necessity we understand, can discern via physics and logic, and can exhaustively seek out (or create) as clearly separate from the rest of the multitude of aspects of our reality as we first encounter it. From a Jacobian

[62] Tegmark, Multiverse hierarchy, 102. Tegmark has four tiers of infinite multiverses, all of which are real, in his view, in some manner.

[63] Alan Guth, David Kaiser, and Yasunori Nomura, 2014, Inflationary paradigm after Planck 2013, *Physics Letters B*, 733: 115.

perspective, the position that the infinite multiverse is really real is clearly an internalist belief geared toward sustaining the fiction that world-building based on a principle of necessity is a coherent activity for Nature to carry out (and for us to track or reconstruct precisely via scientific observation or rational speculation), one generative of an outcome that can really "be" the world. The infinite multiverse is a *nihil* for a Jacobian from the outset; the whole model seems to quite clearly reduce to an identity akin to "infinite necessity = infinite necessity," whatever [the oxymoronic pairing] "infinite necessity" could even be said to mean. It seems that the infinite multiverse must at last disprove the efficacy of using necessity as an ontological "generator" even for the cosmologists: calling the presence of infinite copies or versions of anything, explanatory of the presence or being of any one of them, seems worse than illogical. As Ellis writes, "multiverses allow anything to occur[64]". This seems the antithesis also of what science is good at, and aims for: to extract consistent, testable, "necessary" rules from, and about, and "in", nature, allowing for precise predictability about various of its behaviors. If in the process of saving Nature-as-necessity, we must render our own universe so far into sheerly necessitarian terms and at such an absolute scale that it becomes only chaotic, a less-than-infinitesimal fully subsumed into post-formal chaos of a meta-totality type, it is surprising that cosmologists continue to accept it or do not at least consider how it appears when a Jacobian mirror is held up to it.

Yet the "anything that can happen will happen an infinite number of times" (cf. quotation from Guth above) trope remains an accepted formulation of the infinite multiverse, and the infinite multiverse in turn remains generally accepted by at least a significant portion of cosmologists. Tegmark writes, "in an infinite space created by inflation, everything that can happen according to the laws of physics does happen. And it happens an infinite number of times[65]"; Greene, that in the multiverse things have infinite numbers of copies of themselves and so does every aspect of nature.[66] And by around the turn of the twenty-first century[67] excursions into multiverse theory took on more of an expository than a questioning character, while cosmological discourse even turned to a new topic (as though the other were solved): the place of mind in the universe.

[64] G. F. R. Ellis, The domain of cosmology, 29.

[65] Tegmark, *Our Mathematical*, 123.

[66] Greene, *Hidden Reality*, 39f.

[67] This shift is remarked upon by Rees in Rees, *On the Future*, 187–188.

Mind trying to think its own history and place as such in the universe becomes referred to as the fifth "Copernican" shift cosmologists seek (the infinite multiverse being the imagined fourth). Per above, cosmologist John Wheeler pursued, in roughly the 1970s and 1980s, an especially radical solution to integrating mind more fully with the theory of the universe: he proposed that human beings qua observers *actually create* the universe, per above. For Wheeler, with his participatory AP, the creation of the universe is always ongoing. The human observer engenders "a process of creation that can and does operate anywhere, that is more basic than particles or fields or spacetime geometry themselves, a process that reveals and yet hides itself[68]". This is Wheeler's "it from bit" information view of nature, in which he embraces several arguably neo-German-romantic starting points – no substratum of being which makes the physical world, but only the mind in nature, able to know (and make) it directly; as he puts it, "no tower of turtles", i.e., no "infinite [causal] regress" – and no blueprint for the universe in advance: in his words, no "structure, no plan of organization[69]", though necessity-relations remain paramount within, and as the driver of, what unfolds as the universe. This resembles complete freedom for Nature as ideal and carried entirely by mind, limited only by mind's own creativity (in romantic fashion); mind, for its part, can now finally literally occupy the complete standpoint of Nature's production of the universe, as mind now literally, physically and completely takes over the function of making what is.

Wheeler holds that only a "loop" exists, wherein physics causes observer-participancy, observer-participancy causes information, and information causes physics.[70] In neo-romantic fashion he writes that at the big bang, there were "no gears and pinions, no Swiss watch-makers ... not even a pre-existing plan"; he further holds that physics is as "foundation-

[68] John A. Wheeler, 1982, Bohr, Einstein, and the strange lesson of the quantum, in *Mind in Nature: Nobel Conference XVII*, ed. R. Q. Elvee, San Francisco: Harper & Row, 17.

Note the implication that Wheeler is referring to *nature* through the tacit reference to "nature loves to hide" (*phusis kruptesthai philei*), the famous epigram by Heraclitus, which can also be translated to evoke the sense of an inner essence of a thing now being made visible, then the next moment becoming invisible again; cf. on this epigram Pierre Hadot, 2006, *The Veil of Isis: An essay on the history of the idea of nature*, Cambridge: Harvard University Press, ixff., 1ff.

[69] John A. Wheeler, 1999, Information, physics, quantum: The search for links, in *Feynman and computation: Exploring the limits of computers*, ed. A. J. G. Hey, Reading: Perseus, 313.

[70] Wheeler, Information, 313–314.

free as a logic loop", and "what we call existence is an information-theoretic entity[71]". Conflation of the human mind with the real universe is, in typical romantic fashion, deliberately celebrated as a virtue of the system: the universe is proactively claimed to be mind, and even more particularly, "information", implying science and number as not just its most appropriate language of description, but even the actual direct carriers and real generators of its real smallest components or "bits". In a universe that is only information, any correct scientific utterance "is" reality and in sum total can "be" the universe. In the information view of the universe, necessity truly reduces to logical correctness, but logical correctness can now finally and fully imply [informatically-derived] being; information, an even less restrictive requirement than that of epistemological correctness, is now one with ontological fact. Logical analysis, deduction, or induction, can now literally make and invent the romantics' quadratic "nature" in an almost magical or conjuring sense, evoking Novalis' "magical realism".

Wheeler further expresses a desire to find a "super-Copernican principle" which (reminiscent of Hartle's notion, cf. above) would reject "now-centeredness in any account of existence" and rely in part on observer-participants of the future to help continue building the universe.[72]

Wheeler helped lead cosmology into a speculative trend on how to better integrate or address mind, a trend which has grown in the early twenty-first century. Davies writes of his interest in the "loop of existence that links the highest organizational level (mind) back to the lowest level (particles and fields of matter)[73]". Smolin explores the (also itself neo-romantic and neo-Spinozan) theory of panpsychism, the idea that all matter has some level of consciousness or potential consciousness, speculating that perhaps "consciousness is a general intrinsic property of matter[74]". Carr writes of how science will now need to include "more explicit reference to *consciousness*" and sides with those other cosmologists like Penrose and Linde who would be "skeptical of claims to be close to a TOE [Theory of Everything], when such a conspicuous aspect of the world is neglected" as *experience,* adding that "in the last few decades there have been several hints from physics itself (e.g. the Anthropic Principle) that mind may be

[71] Wheeler, Information, 314, 313.
[72] Wheeler, Information, 318f.
[73] Paul Davies, Teleology without teleology, 106.
[74] Smolin and Unger, *Singular Universe,* 483.

fundamental rather than *incidental* features [sic] of the Universe[75]". Epistemology and ontology – the contours of how we may reliably know things and the contours of what is really claimed to be or exist – were always assumed to be interchangeable in CM, and now the new directions in considering the world to only be information we can manipulate fuses them ever more tightly. Proposing information as being seemed to allow, at least for a time, the CM ontology to pass through the self-biting juncture of the ouroborous presented by the infinite multiverse and, ignoring the indications the multiverse so blatantly offered that the overall sense of a world "made" by necessity was in question, begin again around the circuit, following along a smaller coil within the first.

Big Bang Cosmology's New Mythology of Reason

The big bang cosmologists round out their self-interpretation of their quadratic nature, and of the act of pursuing knowledge of it, with a new version of an early German romantic new mythology of reason project. Their version is far less nuanced than the German one (the German one having been developed, after all, by thinkers who were of the poetic and philosophical bent rather than the quantitative one), including by being far more based in the present-tense, in contradistinction to the romantic promise of a *future* new mythos of reason in an age to come. Sometimes the new mythology regime of the cosmologists tries to take on a patent neo-romantic metaphor or language. Cosmologist Carroll claims to be working on a new "poetic naturalism" based on his scientific naturalism; while for Carroll, naturalism does require that there be "only one world, the natural world, exhibiting patterns we call the 'laws of nature,' and … discoverable by the methods of science and empirical investigation", in which "teleology or transcendent purpose" must be ruled out, there is room in his *poetic* naturalism for values to be created by people, including values of goodness and meaning.[76] In an essay on "Natural Law and Freedom", Carroll puts forward a view of how the efficacy of physics to explain a self-contained world does not rule out human freedom on

[75] Bernard Carr, 2017, Black holes, cosmology and the passage of time, in *The Philosophy of Cosmology*, 61.

[76] Sean Carroll, 2016, *The Big Picture: On the origins of life, meaning and the universe itself*, London: Oneworld Publications, 11, 15ff., 21.

the macroscopic scale of daily events.[77] For Rovelli, when we grasp the "elementary weave of the world" we gain "the breathtaking beauty of the panorama of reality that can be seen from this perspective[78]", using the metaphor of a view to refer to an inner ecstasy of contact with the one Nature-nature as a unity through correct (and ever 'deeper' or 'higher') knowledge of it. This approach also evokes the German romantic Alexander von Humboldt's multi-volume *Kosmos,* mentioned above, which helped introduce the idea of a cosmic universal natural history which included the rise of the human. Humboldt aimed to offer the lay public the sciences' "insight into the connection of the vital forces of the universe" since it is the "noblest fruit of human civilization … the highest point to which the most perfect development of the intellect can attain" (*Cosmos,* II/467).

Another element of the cosmologists' new mythology of reason program is the implicit goal of using our scientific insights into the universe for the purposes of social improvement. Pagels holds that the cosmological and physical sciences are Copernican and objective, but in a somewhat Wheelerian (and German romantic) fashion also thinks they offer a code which can be useful in "programming human social and economic development[79]"; "The challenge is to bring these invisible realms [of the worlds of molecules, atoms, and the endless reaches of space and time] to consciousness and to make human the enormous powers we find there[80]". This evokes a normative desire to harness scientific insights into the way laws "cause" or "make" reality to both re-enchant and materially improve human existence.

There is also recurrence of the romantic new mythology of reason approach in cosmologists' tendency to ironically rationalize the traditional creation myth using their theory, offering big bang cosmology as a "creation from nothing" account somehow "like", but also superceding any continued belief in, traditional creation myths such as the Genesis account. This occurs in contexts spanning the entire cosmological discourse, from the most technical, where mythological metaphors or asides pepper the reasoning, to the most popularized, where the authors take liberties with

[77] Sean Carroll, Purpose, Freedom, 307–308. Krauss also hints in this direction, cf. Krauss, *Universe From Nothing,* 143–144.

[78] Carlo Rovelli, 2016, *Reality is Not what it Seems: The journey to quantum gravity, trans.* S. Carnell and E. Segre, New York: Riverhead Books, 8.

[79] Heinz Pagels, 1982, *The Cosmic Code: Quantum physics as the language of nature,* New York: Simon & Schuster, 346–347.

[80] Pagels, *Cosmic Code,* 348.

directly comparing the big bang narrative to a "myth" with social significance. The neo-romantic claim is, as it was for the German romantics, one of both continuity with "old" mythology and its complete overcoming; ending it as a genre since the "new" mythology meant that the "old" was no longer needed, and was no longer possible. The ability of the poet-philosopher, now the poet-scientist, to combine mythical terms with rationality is used as a sign that the new understanding of nature has permanently displaced traditional myth, and in effect that scientific reason – now that it showed it could be hyper-fluent in its approach to reality, weaving together art, religious feeling, and scientific reason in one omnicompetent view – must now be recognized as the dominant argot in any new [pseudo-]myth.[81]

Zakariya writes that by the 1970s, "prominent scientists located in significantly different disciplinary positions were paying attention to and advancing the idea that a new totalizing myth was at hand" in big bang cosmology's account of universal history; and by the early 21st century it was a widely held belief that "the answers that science can provide to questions about the origins, ends, and dynamics of the universe are all-important" and that science now had the "ability to answer questions previously answered in essentially mythic terms, whether metaphysical, spiritual, or religious[82]". While there were tensions, including in the oxymoronic idea of a "new", and "human-authored", myth, nonetheless the "scientific epic" "began to define what a contemporary myth could and had to be[83]". Despite J. Merleau-Ponty's dismissive response to these efforts (cf. Chap. 5 below) being the normal one among experts on mythology [84], and the fact that Theisen's critique of the original German

[81] This notion is also discussed by Zakariya in the context of big bang cosmologists, though he uses the broader notion of "story" rather than "myth", discussing the scientists' efforts to "finalize" both the content and the *form* of story itself (at least with respect to the natural universe). Cf. Nasser Zakariya, 2017, *A Final Story: Science, myth & beginnings,* Chicago: University of Chicago Press, *passim.*

[82] Zakariya, Final Story, 307 and 9.

[83] Zakariya, Final Story, 337.

[84] Cf. Karen Armstrong, 2005, *A Short History of Myth*, Edinburgh: Canongate, 73–74. Armstrong writes that a creation myth, e.g., "never provided … factual information… [its] purpose was … primarily therapeutic," to "tap into the timeless energies that supported human existence." Thrower criticizes any failure to see mythical ways of regarding nature as simply possessing "an entirely different apprehension of reality from that which now characterizes the predominant Western approach to the world", and opposes the effort to see them as "proto-philosophy or proto-science." James Thrower, 1980, *The Alternative Tradition: Religion and the rejection of religion in the ancient world,* The Hague: Mouton, pp. 17–18.

romantic new mythology project (that the romantics misread "old" mythology via un unwarranted "structural reconceptualization of antiquity under the aegis of modernity[85]") applies doubly in these cosmological cases, cosmologists frequently include a survey of various "old" myths, usually so-called creation myths, in their [especially popularized] efforts to explicate and interpret what the big bang model means, placing the big bang creation story into this grouping as its culmination.[86] The cosmologists not only promise a new holism, a new way of knowing and understanding nature (as, again, implicitly made by or emanating from Nature imagined solely as necessity) which will provide new meaning to human existence, but also heroize the journey to learn the model and so approach this Nature in its purest and most infinite form, by extension indicating that science can and even must become our primary means of romanticizing, or bringing meaning to, our own human existence.

The cosmologists also call on artists and even spiritual writers to join with science to further develop the new mythos. "Contrary to many popular perceptions," Krauss writes, "this scientific story also encompasses poetry and a deep spirituality[87];" this refers not to nature but to the experience of science, again in romantic fashion emphasizing the affective depth of knowledge as a stand-in for other forms of experience we might otherwise value more. The physical world is only the scene which calls the heroic tale of science forth. Krauss romantically equates physics with *other* means by which we can also "change our view of ourselves and our place in the universe" – "great music, great literature, great art[88]". Krauss' work, like most in the popular cosmology genre, is essentially a layperson's textbook on the history of science and the rudiments of the scientific theories of gravity, electromagnetism, and the big bang, interspersed with disconnected excerpts from older traditional myths and sweeping historical claims to have penetrated the logic or "nature" of most human epochs, as well as the physical world, from the ["Copernican"] vantage of physics.

Feyerabend echoes this point, too; cf. Paul Feyerabend, 2016, *Philosophy of Nature*, ed. Helmut Heit & Eric Oberheim, trans. Dorothea Lotter & Andrew Cross, Cambridge: Polity Press, pp. 89–90.

85 Theisen, Romantic Myths, 12.

86 Cf. e.g. Marcelo Gleiser, *Dancing Universe*.

87 L. Krauss, 2017, *The Greatest Story ever Told – So far: Why are we here?*, New York: Atria, 2.

88 Krauss, *Greatest Story*, 274.

Critical Monism's Other Unopposable New Forms

The foregoing CM cosmology matters so acutely in the context of the early 21st -century return of the Controversy because cosmology also becomes the science underlying all others; as we have noted, the CM view insists that the entire world really is only what natural laws have ostensibly made through some imagined algorithmic unrolling of pure law as such, according to necessity. No other accounts beyond science's are used in the early twenty-first century in serious, truth-based discourse on the origin and constitutive essence of the natural world. Philosophical consideration of nature has been subsumed by the sciences' approach to nature since the era of the second Controversy, something which Jacobi seemed to already foresee transpiring in his own day. By the time of the second Controversy, as noted in Chap. 3, theology was arguably already the only vehicle remaining for Jacobian objections to belief in Spinozistic monism's "nature" as really being all of being. CM scientific naturalism, as of the early twenty-first century, has totalized and "finalized" (in Zakariya's term, per above) the intellectual and cognitive landscape even further, leaving very little ground at all where the metarational position can attempt a stand. Even theology now largely embraces scientific world-building through necessity relations in the mode of big bang cosmology, and seeks mostly to offer theological commentary upon this cosmos [or, in Jacobi's view, this empty nihil of reason], whether in a natural theological, panentheistic, or pantheistic spirit.[89]

[89] In the early decades of big bang cosmology, natural theology approaches usually sought to accept the physical description of the origin of the universe according to the big bang, but cite the model's inability to claim ultimacy and completeness, retaining for theology any questions related to the moral or spiritual significance of the entity thus described (and also its ultimate origin beyond the model); cf., e.g., William R. Stoeger, 2010, The Big Bang, quantum cosmology and creatio ex nihilo, in *Creation and the God of Abraham,* eds. D. Burrell, C. Cogliati, J. Soskice, and W. Stoeger, Cambridge: Cambridge University Press; or Stanley L. Jaki, 1989, *God and the Cosmologists,* Edinburgh: Scottish Academic Press.

This division of the spoils approach seemed to exhaust itself with time as scientific cosmology declared itself ever more complete and monolithic, and by the close of the twentieth century and beginning of the twenty-first, many theologians increasingly leaned toward panentheistic compatibilism or integration with the science, as the Spinozistic, romantic moment also swept into theology. Brierley suggested in 2004 that there is a contemporary "panentheistic turn" in theology, led by science; we can read this as a parallel in theology with what Ellerman claimed (see Footnote 102 below) on the philosophy and environmental studies side – i.e., that the present "theoretical moment" is a romantic one. Cf. Michael

Beyond physics and cosmology, philosophy, and even theology, the CM view of reality and other persistent strands of neo-romantic views of nature also influenced, in the twentieth century, the development of systems theory, including complexity or emergent causation theories, and, in the twentieth and twenty-first centuries, the still-ascendent "information" view of the universe. These theories first began to come into their own around the same time, the 1960s, that big bang cosmology was becoming formalized, and would in turn exert their own influence upon it (e.g. via the work of Wheeler or I. Prigogine). These fields also drew from 'post-classical' turns in biology, and then of course after the rise of computing and supercomputing, continued to develop under their influences as well.[90] Key to such fields is the idea of "self-organization ... [which refers] to the emergence of stable patterns through autonomous and self-reinforcing dynamics at the micro-level[91]". This is of obvious proximity to the ideas of nature explored by our four romantic thinkers, especially Schelling; it is also a return of what Jacobi would call the "weaving of weaving" view of the world which is always effectively identitarian as a real *ontology* of the world itself, regardless of which interventions and technologies it permits us to build. Levinson notes that fields like "general systems theory, computational philosophy, artificial intelligence, non-linear and complexity theory, self-organization theory, and cognitive science" with their "paradigm of dynamic materialism" are not congruent with any paradigms of "nature" *except* Spinoza's (though I would say a Spinozist's).[92] Gare cites various late twentieth century "theories of emergence" such as "hierarchy theory, post-reductionist evolutionary theory,

Brierley, 2004, Naming a quiet revolution: The panentheistic turn in modern theology, in Clayton & Peacocke, *In Whom We Live and Move,* 1–15.

Stannard, Peacocke, and the cosmologist Paul Davies are all examples of this panentheistic compatibilist position, all interpreting (in varying ways) the cosmological universe *as defined by big bang cosmology,* whole-cloth, as a kind of *de facto* complete[d] physical created world, and as itself part of God. Cf. Davies, Teleology without Teleology; R. Stannard, 2004, God in and beyond Space and Time, in *In Whom We Live and Move,* 109–120; and A. Peacocke, 2004, Articulating God's presence in and to the world unveiled by the sciences, in *In Whom We Live and Move,* 137–154.

[90] M. Levinson, 2007, A Motion and a Spirit: romancing Spinoza, *Studies in Romanticism,* 46: 367–408, 559.

[91] D. Anzola, P. Barbrook-Johnson, & J. I. Cano, 2017, Self-organization and social science, *Computational and Mathematical Organization Theory,* 23: 222.

[92] Levinson, A Motion and a Spirit, 374, 400. She is citing in inner quotes Ludwig von Bertalanffy, 1968, *General System Theory: Foundations, Development, Applications,* New

geography and ecology", and also "the study of complex adaptive systems" as all heavily indebted to Schelling and his successors, who aimed to develop a "'morphogenetic mathematics' which would 'be able to decipher the internal, dynamic structure of nature' ... and account for the emergence of patterns[93]". As Anzola et al. note, the complexity and self-organization approach is being used in the social sciences as well,[94] again pointing to early German romantic conceptions of integrating all of life's manifestations under a single unity, and evoking Pagels (cf. above).

Interestingly, twentieth- and twenty-first century environmental studies writ large, across the humanities, social sciences, and natural sciences, also embrace this approach. Environmental studies are in a sense doubly romantically influenced; they assume the same CM universe, drawn from physics and cosmology (and chemistry, biology, and other sciences "atop" these), as the rest of the academic discourse, and take this as the "nature" they must seek to understand and then, on the policy and environmental crisis-response side, protect; at the same time, environmental studies and, on the policy or political side, environmentalism, themselves have their own direct German romantic (and broader romantic) lineage. Arne Naess, the founder of the so-called Deep Ecology philosophical-environmental movement in the 1970s, was loosely a Spinozan.[95] Hinchman and Hinchman note that the entire modern environmental movement can be said to descend from "the Romantic movement and associated currents of German idealism and American transcendentalism[96]".

This inheritance leads the environmental studies approach to not only assume the monistic scientific universe as nature, but also to assume that theoretical and artistic approaches to nature which preserve or build upon romantic thought are automatically conducive to its preservation or possessed of a normatively positive environmental ethical orientation.

York: G. Braziller, 37, by way of Mark C. Taylor, 2001, *The Moment of Complexity: Emerging Network Culture*, Chicago: University of Chicago Press, 140.

[93] A. Gare, 2011, From Kant to Schelling to process metaphysics: on the way to ecological civilization, *Cosmos and History: The Journal of Natural and Social Philosophy*, 7: 67–68.

[94] Anzola et al., Self-organization.

[95] Cf. Arne Naess, 2016, *Ecology of Wisdom*, eds. A. Drengson and B. Devall, London: Penguin Random House.

[96] Lewis P. Hinchman and Sandra K. Hinchman, 2007, What we owe the Romantics, *Environmental Values*, 16: 334. They cite Oelschlaeger, Roszak, Glacken, Hargrove, and Nash as examples of important [Western] environmental thinkers of the twentieth century who derived inspiration and foundations from romanticism.

Schimelpfenig calls on romanticism, deep ecology and something he calls "Renaissance science" as relevant for addressing the environmental crisis.[97] Borgmann holds that the "fundamental [moral] cosmological predicament is much the same today as it was at the turn from the eighteenth to the nineteenth century[98]" and calls for more work on trying to elicit a response in the mode of romanticism to this crisis using big bang science, citing as a positive example poet Robert Pack's "cycle of thirty-one poems that ponder and respond to cosmologist and physicist Heinz R. Pagels' *The Cosmic Code*," uniting "scientific lawfulness with the poetic presence of nature and culture[99]". Philosopher and romanticism scholar Gare makes a case for a romanticizing of scientific cosmology to provide humankind with a source of ethical orientation in the cosmos, including in light of the environmental crisis. He claims Schelling's philosophy offers a hopeful path toward "transcending scientific materialism," "overcoming the opposition between science and the humanities," and "enabling people to understand themselves as culturally formed, socially situated, creative participants within nature[100]".

The early twenty-first century philosophical "speculative realists", interested in reviving philosophy of nature in roughly the German romantic tradition, seek to update it by incorporating the findings of physics and cosmology while retaining, apparently, the essential [neo-]romantic quadratic version of "nature" and showing that thought is capable of direct speculation on *this* nature itself. They reject the Kantian turn (though still seek to argue against and account for it, which arguably still earns them the "critical" qualifier on their CM); Meillassoux calls Kant's limiting of direct speculative knowledge of nature a downgrading of philosophy of nature to mere "correlationism",[101] which we must know we can exceed; this, too, echoes German romantic confidence in intuiting nature as such

[97] R. Schimelpfenig, 2017, The Drama of the Anthropocene: Can Deep Ecology, Romanticism, and Renaissance Science Rebalance Nature and Culture?", *Am J Econ Sociol*, 76: 821–1081, e.g. 821.

[98] A. Borgmann, 2006, Broken Symmetries: The romantic search for a moral cosmology, in *Philosophical Romanticism*, 241.

[99] Borgmann, Broken Symmetries, 258. Cf. e.g. also Matthews for a related, though more generic (less cosmological), form of the same argument. Bruce Matthews, 2014, The new mythology: Romanticism between religion and humanism, in *The Relevance of Romanticism*, 216–217.

[100] Gare, From Kant to Schelling, 68–69.

[101] Quentin Meillassoux, 2011, *After Finitude: an essay on the necessity of contingency*, trans. Ray Brassier, London: Continuum, e.g., 5.

"without mediation" (cf. Chaps. 1 and 2 above) while still linking the products of such intuition with the demands of the theoretical understanding and its need for relations of necessity to "make" the real world; and echoes the interwar logical empiricists' claims to have found, in mathematical physics and our understanding of it, the only possible pathways to ontological, real, true speech about the external world. Again echoing the Vienna Circle, Meillassoux argues that contra Kant, we do have a firm foundation to speculate directly on nature as it is in itself. For Meillassoux, per Badiou's gloss, this is because we know "one thing that is absolutely necessary: that the laws of nature are contingent[102]". The basic effort of the speculative realists is to return philosophical speculation to considering nature itself within an absolutized CM framework where the intelligible universe is a set of all possibles "made" by the operations (or demands) of [absolute] necessity. Ellermann claimed in 2015 that these speculative realists' work helps point us toward "how nearly our present theoretical moment hews to a certain romanticism.[103]" New romantic currents have emerged in other environmentally-inclined twenty-first century philosophical approaches as well. As Josephson-Storm (disapprovingly) points out, the German romantics' "disenchantment thesis" has recently found "fresh purchase" with a "host of thinkers" charging "modern philosophy with despiritualizing nature and rendering matter dead and inanimate," and in response aiming to "recover alternate animating ontologies sometimes referred to as 'agential realism' or 'enchanted materialism[104]'."

A final relevant way in which the CM view of the world as really being "made" through operations of logical necessity has become entrenched is in the beginning of the proliferation of imitation "worlds" which are held to instantiate their own impersonal and necessitarian order (which must be followed by the human subject) at the societal and individual scale. This proliferation, which lately resembles a rising tide of separate worlds or even entire separate rationalities (each with their own true and false standards even in areas previously considered factual and universal), has been exponentially accelerated by the arrival of computers and the internet. But its roots, like those of the information age itself, reach down to the era of the first Controversy and the romantic pathway out of it the West chose.

[102] Alain Badiou, 2011, Preface, in Meillassoux, *After Finitude*, vii.

[103] Greg Ellermann, 2015, Speculative Romanticism, *SubStance*, 44: 154.

[104] Jason Josephson-Storm, 2018, *The myth of disenchantment: Magic, modernity, and the birth of the human sciences*, Chicago: University of Chicago Press, 5–6.

The universe ontologized through sheer logical necessity – what Jacobi memorably described (cf. Chaps. 1 and 2) as the "weaving of weaving" kind of world our theoretical understanding stitches together – turns out to have a highly reproducible form; it is very imitable[105]. We see this in the way necessity attacks itself, in the infinite multiverse: necessity is [in one form] a type of process, and we can "necessitize" any entity by seeking to make of it a perfectly necessary, perfectly infinite, absolute maximal or absolute minimal case of itself. It is a type of rational operation we know how to perform. Jacobi seemed to intuit this self-perpetuating quality of defining the world through necessity-relations. He argued, per above, essentially that to make a "world" only requires that the subject first seize "his life by the root so to speak, in order to throw it from him[106]", then remove everything from his awareness that is real, and then he is free to begin taking apart and putting back together in new forms his founding identity-statement and any rational operations of necessity he wishes to perform upon it. In the simplest methodological terms, once the subject embarks on this exercise, he is free to begin to world-build using any axiom or entity as his founding principle, and then direct necessitarian logic entirely into its service, as it were. Necessity statements, the "truths", or undeniable directives for the subject, this world then frames, follow one typology: "X cannot be otherwise than in accordance with the founding directive of this 'world'". Ontologies of necessity at their root are programmatic and self-contained. They refuse any Thou or externality which would "break in" (in Barth's term) upon their monist structure.

While scientific forms of cosmology, romantic in their core assumptions, may still orient themselves to the best of their ability toward ontologizing nature as generally as possible using only natural laws and claims about matter-energy and gravity opened out to infinite (or perfectly generic) imagined cases, and still attempt to incorporate data and observations, extrapolations from physical systems on earth as they understand them, and so forth, they do this through, as it were, historical, legacy impulses alone. Once neo-romantic science and philosophy elided the three kinds of necessity, and simply posited necessity-relations as sufficient

[105] Imitable and also self-conveying; transposable; while there is not room to go into the evidence for this here, this fact arguably lies at the heart of why romantic monism laid the groundwork for the computing and information age.

[106] This is from numbered paragraph no. 316 of a Jacobi text not strictly speaking usually included in the Controversy texts, F. H. Jacobi, 2016, Edward Allwill's Collection of Letters, in Jacobi & di Giovanni, *Main Writings*, 495.

for making the real (though they dressed it up and monistically defended it by trying to present necessity as an actual trait possessed by a real, infinite entity called Nature), the logic of the information age – which only means, converting and destroying real life and real art by rendering it all into a not-even infinitesimal, just an undifferentiated, member of a single, all-enveloping post-formal chaos – was already showing itself. (We can recall, again, that the romantic understanding, of nature and knowledge influenced Ada Lovelace, one of the first computer programmers, per Chap. 3, and that big bang cosmology now declares that its ostensible fifth "Copernican revolution" has revealed that the universe is really only information, cf. this chapter).

The combination of the ontology of necessity as the only path to knowing what is, and the ability to convert the world into information, to vanish the realm of experience and existence altogether, has as of the early 21st century accelerated the always-latent proliferation of "worlds," each in effect with their own programming and sets of necessity-relations which the subject must mesh themselves with as just one more piece of information or part of a necessity-relation, if it wishes to live within them. This proliferation of rationality-worlds began first within the modern university, where it has also continued to accelerate, so that over time the different fields which all want to mirror science (and thus base their knowledge on comparing some necessity-relations measurable and testable as true or false, against other versions of the same web of relations) have increasingly become hyper-specialized, completely self-referential pods, unable to understand one another, as all the communal metarational bridges which may have survived from past ages gradually disappear. The neo-romantic twenty-first century CM era wants to "unify" the fields (or at least praises interdisciplinarity) but has only done so methodologically in this sense of holding up science (and thus necessity-relations) as the only template for serious academic discourse; necessity-bound worlds as the only worlds. Cross-domain communication and truth-sharing in terms of these fields' content is increasingly impossible, since each field uses a different set of premises as a standard set of truths to dismantle and reassemble in this manner, and again, metarational subjectivity is nearly entirely absent.

Alongside the splintering of rationality in the academy came the rise to ever more dominance over society of another type of necessity-based ontology equipped with the means of force required to make its own world true and real: the capitalist economic system. This system's sole axiom, as Jappe notes, is the pursuit of the increase of "production of exchange-value, which is represented in money[107]", creating a necessity-web with just one main necessity relation to which real life, now, not only real knowledge and truth, must be lashed: this single programming instruction is that it is a value-raising world; all directives take the form "X cannot be otherwise than in accordance with the need to raise exchange-value." Nothing is permitted to be in such a world which does not accord with this necessitarian "programming", this overarching instruction. This, of course, has unleashed a horde of programmed actors into the world who embody the force of this directive and usually exercise it directly against the environment, which this world cannot see, as it can only recognize one type of truth or reality, that X or Y raises value and so can "be" (or that A or B does not raise value and so cannot be). All the individual and social entities – even the West itself as a whole – programmed in this way unleash a wave of force like a gnawing hunger, a *phagic* force (cf. Chap. 3, above), upon the earth; they carry this one necessity forward which in Jappe's terms is "limitless", "abstract," "empty of content," and able to "destroy all … in its path … on an ever-increasing scale.[108]" Again, this value-raising regime imposes the brunt of its vast *necessity-framed* hunger ultimately on nature, since the "expansion of value cannot take place without … material production" and "material growth," which "by consuming natural resources, ends up consuming the real world.[109]" Like all ontologies of necessity, the "world" made solely from spinning out, and combining, and recombining, various relations of necessity flowing from the axiomatized or absolute starting premise – which also pre-determines the primary terms out of which the world itself is constituted, i.e., in this case, production, increase, money (or exchange-value), and so forth – is "*tautological* – it has no content of its own and generates a dynamic of growth for the sake of growth," since "Growth!" is the essence of the command this world issues, its one arch-principle to which all admissible necessity-relations must refer back, and as Jappe notes, "It is easy to understand that the deep roots of the ecological disaster brought by capitalism are to be found here" given the "indifference" of the system "to the means for achieving" its single imperative.[110] For Jappe, this value-driven

[107] Anselm Jappe, 2023, *The Self-Devouring Society: Capitalism, narcissism, and self-destruction*, trans. Eric-John Russell, Brooklyn: Common Notions: 3.

[108] Jappe, *Self-Devouring*, 1–5.

[109] Jappe, *Self-Devouring*, 12–13.

[110] Jappe, *Self-Devouring*, 12–13.

system is so impersonal it does not even spare those who financially benefit from it most, any more than it does those who benefit from it least; its "logic" is "automated" and inhuman, and none can escape it,[111] bearing out many of the concerns Jacobi had pertaining to the uncanny power of certainty as such (cf. below) and of our theoretical understanding as a whole. We can also see how serving, or surviving within, this economic necessitarian regime requires us to annihilate the real, removing it completely from our cognitive awareness, just as Jacobi warned any theoretical world required us to do if we were to try to [tell ourselves we] really inhabit it, or, more accurately, if we were to decide to try to adhere to or abide by it. The real world cannot exist for "Growth!" because actually seeing the world would either prevent us from entering into the "Growth!" command-space altogether, or equip (and confuse) us with competing truths which do not refer back to its single value-raising imperative. To enter any "ontology" or ostensible "world" of necessity, as Jacobi pointed out, one must forsake the external world as it is in itself, just as one must forsake any competing necessity-regime. The external world becomes, as Jappe notes, solely where we go to service or satisfy the phagic intensity of the "Growth!" command, deflecting its might onto whatever we can extract or make from the world, in order to keep it from devouring us, ourselves (though it often does this too, as per Jappe's overall thesis).

After the scientific and the economic, a third type of world-construction using relations of necessity (and we are not here proposing these types each stand apart; they rather nest one within the other) is also currently proliferating in our CM-dominated West, this one at a smaller, and even often at the individual, scale. Each of these constructed private "worlds" or even rationalities may be small in scope, but overall as a social practice the building of these worlds has already had enormous impact, and these worlds constitute a still arguably vastly-underestimated element of our third Controversy, both in terms of the evidence they provide of reason's ability to become irrational, and the way they repeat and accelerate the ways in which the value-raising economic regime also causes environmental impacts which the subject cannot even see or frame, let alone cease or reverse, because to do so would contravene the necessity-regime within which the subject is being asked to survive (and usually not "by choice" in any classical sense). Through the use of mediating

[111] Jappe, *Self-Devouring*, e.g., 198; or 192 (where he is also citing Andre Gorz, 2010, *Ecologica*, London: Seagull Books, 4.

technologies, and with the availability of an unprecedented amount and type – even a true sheer chaos – of informatic material, which CM increasingly asserts is the same as real being or that which generates reality, conveying to the subject that they need not look for anything beyond the informatic, subjects can increasingly define their own chosen and arbitrary premise as the master necessity of their world to which all statements of truth or necessity (or directives which must be heeded) must refer back. The subject performs this world-creation with perhaps a sense of increasing their own freedom, but by and large, such efforts only serve and sit within the "Growth!" necessity regime, turning personhood and truth themselves into mere fodder to serve its directive. These individual necessity regimes are also humanity- and person-jeopardizing in that they are non-communicable via metarational means (but only through exact repetition or replication), leaving each subject without a firm foundation in how a fuller type of reason operates, and potentially one day even devoid of any sure ability to know truth from falsehood. This may lead to further forms of irrationalism and the abandonment of the truth-claim as such, as well as the abandonment of subjecthood itself, whether in favor of AI-produced avatars or new (and potentially multiple) personas only present online, or, as Jappe raises, the end of individuals as such in favor of a "multividuum" (cf. below). Given how well the necessitarian quality of the new individual regimes, and their service to "Growth!", are currently disguised under the narrative of enhancing individual freedom (or freeing us from our excess baggage, as in Hartle's expression above), it seems it would only require someone figuring out how to raise value and serve "Growth!" via the end of personhood as such, in order for voluntary relinquishment of the self to "take off" as the next trend. AI in a sense offers merely a vivid reminder of the way in which the theoretical understanding, for Jacobi, has always been inhuman[e] when untethered from the fuller self. That serious people discuss AI "taking over" is perhaps only a latent recognition of how deeply, and for how long, the West has been committed to trying to inhabit only an ontology of necessity, making it seem impossible to turn away from anything our "world" rewards as in line with its regime, no matter what this thing is. The half-hearted discussions of how to halt the possibility of a world-ruling AI (as obedient necessity-regime participants, again, it seems far more likely the West would simply accept and even in part welcome its reign, were this actually a realistic possibility) also suggests this technology might even be helping some realize, belatedly, how authoritarian the theoretical faculty based on necessity-relations (and associated commands to the subject) has always been; how brazen and inherently incapable of self-limiting (as Jacobi repeatedly tried to point out). Now this form of reason rewards, and wants to elevate, the sheer computation and mimicry capacities of AI; the theoretical faculty asserts the "artificial" in the title does not matter, that its need for a technological substrate and its lack of a soul or spirit does not matter, since information is the same as being; effectively, and by extension, it asserts that artifice and authenticity are one and that there is no such thing as a real external world outside of the mind (which now only possesses information and necessity relations and directives, not any metarational cognitive ability, and so cannot encounter any world it does not first create in any case). The inhumane theoretical faculty, with its ability to see only necessity relations, has new, powerful tools at its disposal, but in its assertion of primacy over the person; its annihilaton of the real external world; and its banishment of the metarational capacity which could refuse the orders of any necessity regime or insist on preserving its awareness of the real world as a really external Thou, the basic dynamics of the threat the unchained, decontextualized theoretical understanding poses to the subject are effectively the same as they have been since the first Controversy. If we ever so completely relinquish our selves as to desire a fully informatically-created one, the subject will have to bear a final violence which is the death not only of the last surviving shards of the metarational person but finally the subject as a whole. Neither faculty of reason could continue to function in such a case. As Jappe notes, some believe in the future individuals will be subsumed into a "multividuum" where the "individual human being would only exist as an element of a collective intelligence, comparable to bees … This intelligence would reside in the cloud.[112]" Even the axioms and founding necessities from which we make our most rationalized ontologies seem unlikely to be able to survive such an arrangement; reason would again collapse altogether. Even if it clung on, in any of these arrangements, as with the intensely phagic array which keeps the economic (and even some parts of the scientific) ontology of necessity intact-for-us, the "multividuum" itself would only be able to last for as long as the actual external world, our environment, proves able to

[112] Jappe, *Self-Devouring,* 157, also citing Pierre Levy, 1999, *Collective Intelligence: Mankind's emerging world in cyberspace,* Cambridge: Perseus Books; and Ippolita [an Italian writers' collective], 2013, *The Dark Side of Google,* Amsterdam: Institute of Network Culturs.

bear our use of force to literally and physically make the conditions we need to run it.

Overall, it is not difficult to see that for the third time, the Controversy "configuration" (to again use Lazier's term, cf. Chap. 3) has returned, a configuration in which the West must understand and acknowledge that once more, a very large and portentous question mark has been raised as to the "fate of reason" *itself* (in Beiser's phrase). Yet still – seemingly, with each iteration of the Controversy the Spinozistic side has only gained in strength, and thus far this iteration seems to be no exception. There remains barely a sliver of ground in, or even adjacent to, the mainstream Western discourse on which to take up a neo-Jacobian stand. Nowhere remains to make such an oppositional, not refutational[113] stand save the bare ground of the rapidly degrading and vanishing non-human world itself, the very place that suggests a neo-Jacobian opposition must, no matter what, be at least attempted. The Jacobian position must be a rural one and not an urban, urbane one; its argument must be heard outdoors. This external reality and what humans are actually, physically causing there is where we will turn next, though conditions there likely do not need much elaboration for the twenty-first century reader attentive to their surroundings. After a brief sketch of these conditions, we will move in the final chapter to elucidating further the beginnings of a neo-Jacobian response to both the human (which we have just reviewed) and the non-human (which we will next briefly review) evidence all around us which indicates that this iteration of the Spinoza Controversy, this crossroads where we must confront an unwelcome either/or where the "fate of reason" hangs in the balance, calls for a different road out of it to be chosen than has been the case twice before.

Humans as Vanishers of the Living World

In the context of the Controversy, the environmental conditions which humans– first and chiefly Westerners but increasingly, all other humans as well – have created immediately testify to the extraordinary power, and the extraordinary consequences, of the type of naturalism which posits the "complex that we call *nature* [114]" (in Jacobi's terms) as more real than the given, really existing world, and increasingly also proliferates additional

[113] Cf. Sandkaulen, Jacobi and Spinoza, 47-48; cf. also Chap. 2 above.

[114] Jacobi, Concerning the Doctrine, 376 (no. 425).

ontologies of necessity which mimic and overlap with this "complex," as we have just been discussing.

Humans have seemingly become fully locked in to imposing technologies of conversion and enclosure that re-make the world according to their ontologies of necessity everywhere they go. These interfaces and systems – upon which the West currently relies for our daily food, water, and shelter, let alone a thousand other needs real and imagined – are, first of all, designed using science and engineering and their view of the way nature works and what nature is. Parasitically atop this runs the monist economic necessity world where value must increase, and the fastest way to do this, is to use the interfaces and systems made by human beings to eke a little "value" out of the system for the individual, in the form of money, every day. Out of both these complexes emerge ever more humans and ever more rationality-systems, increasingly arbitrarily chosen and constructed – in part, ironically, in the effort to escape from the spatial and social effects on human existence of the first two necessity systems being taken as more real than the non-human earth as it is in itself.

While humans, like all biological beings, have always been, and must remain, partly phagic, since we must take in energy and calories to stay alive, historically there has tended to be at least some phanic counterbalance in our perception and use of the external world – whether out of self-preservation alone or for other reasons of simple awareness, interest, respect, awe, religious purposes, or fear. The need to make true, and make be, various ontologies of necessity, each with their embedded necessity relations which the subject must obey and mesh itself into, however, has placed the phagic in full control of the person and their life, banishing the phanic faculty along with all the rest of the metarational subject. The person furthermore increasingly needs a hyper-developed phagic faculty to navigate the mostly domesticated, synthetic, and informatic world in which it increasingly solely finds itself, this world with its various sets of anthropogenic systems which require only logical obeisance from the rational self – the only part of the subject which can understand or mesh with these worlds as they require us to do (these worlds being, again, almost exclusively now the products of solely the theoretical understanding, with its attribution of all things to "necessity" and its obedience to several necessity-regimes in turn). We *make* industrial systems and structures, mega-cities, suburbia, box stores, ceaseless entertainment and internet informatic immersion domes available at all times, automobiles and airplanes, cheap consumer goods and groceries, fast food, a thriving global black market where drugs, weapons, and other human beings are available for purchase any time, into our world; the other, real world vanishes. Wilderness becomes a site of opposition to this enforced world-construction, a place of rediscovery of what really is external. It is perhaps because valuing this kind of site cannot uphold scientific, economic, or subjective necessity as the only law and as all that is, that wilderness has come under increasing attack in recent decades even from within the environmental movement.[115]

Humans are quickly rendering the earth a spatial-aesthetic-experiential cacophony (akin to our own cognitive world, which is being rendered an informatic chaos) due to our various worlds which coalesce in spatial terms in conjoint landscapes of accumulation.[116] Accumulations, heaps, and ruins, places where the entropy has been run up to 1 (at least for the near term), are the spatial result of all of our ontologies of necessity being built and enacted, being believed in, and served, above the external world itself. From the oil and gas extraction and transport infrastructure which fuels and enables most of the material conditions required by all three types of necessitarian world-building, to the industrial dumping-grounds full of e-waste and old cars, to the pollution of rivers and oceans with the chemical and plastic residue of our constant buying, the natural world's actual situation becomes increasingly invisible beneath and behind the light, sound, and interfaces we have erected in its stead spatially, just as we began our world-building by annihilating it cognitively. The risks grow for all other species (and our own considered as just one more biological species) by the day. Seen from the other species' perspective, the human species must be perceived as the senseless killer, with no rhyme, reason, or predictable pattern to its acts, which again are undertaken without even seeing the world as it is in itself, but only (and even here only in the narrowest sense possible) how it can serve to bolster, uphold, or allow us to obey one

[115] Wilderness as an idea and a policy is under attack both from the indigenous people's side: cf. e.g., M.S. Fletcher, R. Hamilton, W. Dressler, and L. Palmer, 2021, Indigenous knowledge and the shackles of wilderness, *PNAS,* 118; and from the neo-greens in Western conservation movements, as is described in, e.g., D. Kidner, 2014, The Conceptual Assassination of Wilderness, in *Keeping the Wild: Against the domestication of Earth,* eds. George Wuerthner, Eileen Crist, and Tom Butler, Washington: Island Press.

[116] Cf. Katherine Snow and Sjoerd Kluiving, 2022, Reading the Anthropocene through Landscape Archeology: Historical and Contemporary Landscapes of Accumulation in the Netherlands and Lebanon, *The International Journal of Environmental Sustainability,* 18: 1–22.

or more of our rubrics of necessity. Habitats which took hundreds of thousands of years to form are clear-cut for soybeans; species which evolved over millions of years are harvested to the point of extinction in the span of a few thousand. The external world is increasingly also not seen because we are vanishing it. We vanish physically what we began living in this way by vanishing cognitively. Subjects meshed within one or all of the three types of ontologies of necessity cannot truly act in a way in line with environmental cognition no matter how they may temporarily try; there are impersonal rules and necessities which override this and do not permit the metarational perspective to survive, let alone direct the fuller person to act against one or all of the rationalized necessity-webs by which the Western subject is increasingly bound and controlled in the belief that they "are" the world(s) they *must* really fully inhabit.

Those corners of science which consider themselves environmental sciences, begin in the same kind of necessitarian world-building but then hope to bring it onto a normatively positive application-track at, as it were, the last minute, and do so while still using necessitarian reference-frames rather than any metarational ones. CM naturalists who hope to "use science" to "protect nature" share an underlying conviction with the majority of people (including those in other parts of the sciences), who never conceive of such a goal as protecting nature, that nature is a knowable quadratic made of necessity relations and not essentially describable any other way. This sets it up as a mere (perfectly rational) disagreement when environmental scientists disagree with oil executives or hedge fund managers over how best to, as it were, ride these rails. How can one argue over what one "should" do to networks of necessity? There is nothing that can be found within such a mesh that could even offer a meaningful choice. Science cannot even be surprised that its own environmental models of climate disaster and species extinction, which seem to threaten the very sensicality of the economic value-raising world, are countered by dubious science-like sources and factitious maneuverings backed by the value-raising system itself. As discussed above, the ontology of necessity has already become profligate; we have now passed the point in the West where there will always be an alternative world on offer at the subjective level, or another, stronger, more violent insistence at the economic level that value-raising and value-raising alone is real and necessary. CM science created this situation; these are the chickens of the CM ontology of necessity come home to roost. CM originated the stream of belief in the power of theoretical reason to world-make and to really know that the world out to its real infinity could be adequately presenced and really ontologized as a system of necessity-relations imagined as bearing the power of being, of life, itself. This stream of belief has become a land-devouring flood with major eddies and whirlpools spinning off of the central current even more dangerous than its crushing, straight, totalizing waters. The notion of world-creation via knowable necessity alone, the main scientific trajectory along which the West has long been moving, ends in the deeps of trying to inhabit a chaotic and arguably senseless "informatic" *nihil* of reason where information is allegedly existent, everything is infinitely repeated, and everything and nothing has meaning, while the real external world disappears altogether. "Environmental" science cannot be surprised that it is being beaten by those who are more at home in the essentially contentless character of all fact-based closed-loop systems than science itself (with its rich legacy inheritance from theology and metaphysics, and concern with empirical observation) – that it is being beaten in effect at a new version of its own game.

Just as the infinite multiverse conception, in seeking a universe made only of necessity and meta-necessity relations somehow unfurling in a self-contained manner which is also absolutely ontological, ends up undermining what meaning can be found in calling a thing (or nature as a whole) necessary, the ontology of necessity built on value-raising must (and can only) eventually end when it completes its own loop and makes the notion of value so generic and infinitely necessary as to be meaningless; when it arrives at a fully environmentally degraded world where only information remains, and all things have the same, extraordinarily small amount of value, making everything valueless. In the interim, no effort to proffer the findings of a competing necessity regime to that of "Growth!" can even register. Of course, the environmental costs of letting the "Growth!" regime run until this point at which the very notion of value is understood as having been destroyed has horrific environmental and subjective consequences. And so, the Controversy continues to present its stark either/or.

For a Jacobian, of course, "environmental" science also cannot mount a response to environmental decline simply because we really *do not inhabit* the sciences' "world." The external world that is declining really is not the same thing as models of nitrogen levels over time in the soil, graphs of species extinctions, or supercomputer models of human population and calorie needs for the next century. The external world does not exist in data on freshwater per capita, fishing rates, dead zones, deforestation, vertebrate species survival, or carbon monoxide emissions, that environmental scientists try to use to stir responses from the public and from leaders.[117] Science, in an extension of the cosmological ontology of necessity, envisions the environment as a mesh of necessity-relations in which they could in theory controllably intervene. But the living earth as the really real place we dwell is not the same thing as such a mesh, even if it in some sense does contain such causal networks (including because we have built and inserted into the world what we consider proof of these networks and regimes, and constantly refer back to our own interventions). All we can do when we try to deploy environmental science to preserve the real world is build some equipment, or carry out a quantitative study, or author a report, using logic and analysis of natural laws and chemical structures, and the behavior of what we call matter, and so forth, to tease out and then recombine elements we ourselves, as Jacobi would point out, had first created and defined.

Biologists highlight an ongoing "sixth mass extinction" on Earth – the first five were natural in origin, with life on Earth recovering only slowly, over "millions of years", while the sixth has come at the hands of modern man; "[H]umanity has come to pose an unprecedented threat to the vast majority of its living companions" as we are the agents of an "ongoing biological annihilation.[118]" Annihilation is an apt term also in the Jacobian sense, as we can understand the annihilation of species as part of our having rendered the external world nonexistent, torn ourselves away from it by the roots, in Jacobi's metaphor (cf. Chap. 2), so we could replace it [cognitively] entirely with our own nihils of theoretical, necessitarian reason. We humans are physically, literally, vanishing the natural world, removing the living layer of the external world around us (and also altering much of its non-living layer), making it so that it is no more, and from the neo-Jacobian perspective this is a predictable correlate of naturalism. The annihilation of other species through "human activities … habitat loss and fragmentation, illegal trade, overexploitation … toxification, and pollution" is perhaps what is causing this "extinction crisis," the "toxification and climate crises," and the "existential threat" not only to all these forms of life other than human beings, but to *human civilization itself*[119]; but

[117] E.g., in W. J. Ripple, et al., 2017, World scientists' warning to humanity: a second notice, *BioScience,* 67: 1026–1027.

[118] G. Ceballos P. R. Ehrlich, and P. H. Raven, 2020, Vertebrates on the brink as indicators of biological annihilation and the sixth mass extinction, *Proceedings of the National Academy of Sciences - PNAS,* 117: 13596.

[119] Ceballos et al., Vertebrates on the Brink, 13600–13601.

what the subject is doing which allows this to all appear rational or necessary, is following the dictates, as it "must," of its substitute worlds *made* sheerly of necessity relations. The physical removal of the world is both the outward manifestation of the removal of the world within the subject and part of what is needed thereafter by the subject in order to sustain the illusion (and/or make it true, through direct conversion or transformation of the real into the domesticated, the synthetic, or the informatic) that the subject's ontology [or ontologies] of necessity (whether scientific, economic, and/or subjective) within which it attempts to survive, gauge truth, and find meaning in human existence is [or are] the same as the really real external world.

Our own vanishing of the non-human realm does not compare with past causes of great loss of life on Earth due to volcanic activity, encounters with asteroids, and so forth. Those others were isolatable events only in the context of geological time: the impacts they caused unfolded in each case over periods of tens of thousands to over a million years.[120] Our destruction and removal of other species, and the landscapes and ecosystems they created over millions of years, is almost unbelievably rapid and can only be sufficiently critiqued by understanding it as the literal correlate and proof of our absolute cognitive nihilism with respect to the real. Jacobi understood that in an internalist, theory-defined world declared to be external, there was no check on our preference for self-worship; he foresaw the nihilism that would result. CM science, the inheritor of romantic confidence in the capacity of human reason to grasp nature as Nature and romantic belief that causal necessity-mapping forms of reason should be the master-language into which our sense of beauty, goodness, and truth (in all its forms) as well as ultimate ontological claims must be able to be translated; the inheritor of the logical empiricist's seamless monistic nature, the vision of which could elevate the human soul, and of the Vienna Circle rejection of any notion that "fidelity to evidence and punctilious argument is somehow small-minded[121]", is arguably wholly incapable of understanding the root of the environmental nihilism problem we face. In line with its own necessity-directives to see nature only via and "as" sets of laws and the necessity relations they establish as valid and ontological, it cannot judge its own frameworks as inherently world-

[120] Olivia Lai, 2021, The World's Mass Extinction Events, Explained, Earth.Org, https://earth.org/what-and-when-were-the-mass-extinction-events/

[121] Cf. Creath, Logical Empiricism, NPN.

annihilating. It can only ever show the Western subject ever more details of its ontologies of necessity and how to deploy them. It rejects the metarational subject's freedom to think and choose whether and when to utilize logical necessity's worlds and when to make decisions grounded in environmental (and other forms of spiritual or complete) cognition. Science can only insist, via the alchemy which Jacobi rejected as impossible, that the actually existing world really is "made" of inevitabilities (necessities) and their playing-out, and is really possessed of generics that act in the world; science can only offer a monistic unity, devoid of real incommensurables. Science cannot see that naturalism as such must have environmental, spatial ramifications because it cannot see them after severing itself from the metarational frame and person; it cannot grasp why the Western romanticization of theory has led to a complete failure of environmental cognition because since the time of the first Controversy it has thrown its lot in with this very romanticization.

There are many ways, again, of describing the environmental condition the West has made globally, but in the framework of the Controversy it appears as the very spatial correlate of the nihilism of which Jacobi warned more than two centuries ago. What the scientist says, that we are the agents of an "ongoing biological annihilation[122]", is true, but we need Jacobi's understanding of the true source (and extent) of the *nihil* in the middle of that word, annihilation, order to properly understand and actually respond adequately to this true, and incomprehensibly serious, accusation. And the nihil at the center of this word must be examined and debated in these radical, Jacobian versus romantic, terms, if there is to be any chance that the West reinstate the metarational subject and begin to move toward a capacity for Jacobian environmental cognition. The West's environmental discourse must be made far broader and deeper, beginning with considering both sides of the Controversy afresh in the environmental key I am seeking to begin to lay out as possible in this work, and will hopefully continue to elaborate upon in the follow-up volume. We are in a third Controversy because it must be *at least* a crisis of and for *theoretical, logical reason itself as well as of and for our fuller sense of reason* if we have come to such a precipice as the deliberate "vanishing" by human societies of so much of reality. If we do not take to the bare hillside now, the non-human realm, the external world itself, will eventually, of course, have the last word; but this word will not be survivable by most species,

[122] Ceballos et al., Vertebrates on the Brink, 13596.

including perhaps our own, even if it is later succeeded by a whole new round of evolution our species will never witness – an ultimate contrast indeed with the future claimed for humankind by the new mythology of reason.

If the first two iterations of the Controversy saw the West embrace the Spinozistic side over the Jacobian one both times, twice rejecting metarational critique of our ardent belief in the ouroboros-world of our own making, the added evidence of one more century in which our devouring and annihilating of the world has reached unprecedented levels – while at the same time the ontologies of necessity which order it to happen themselves increasingly reveal their irrationality – must now elicit a third, final re-arbitration. Due to our vanishing of the real around us, we now have quite literally, to use Jacobi's phrase, only a "nothing everywhere manifest," or a "manifest nothing,[123]" left to lose.

[123] Jacobi, On the Divine Things, 174.

CHAPTER 5

Environmental Nihilism as the Limit of Reason

The Spinoza Controversy, in my reading, ultimately turns on whether theoretical reason will be controlled and directed by the whole, metarational person, by the inner self, or whether the reverse will hold. Any neo-Jacobian opposition to the annihilation of the external world as the spatial correlate of CM originates in the restoration and preservation of the metarational subject. Before the subject (and all other life on earth) can regain the freedom to live in and experience a world not supersaturated by the activities and outputs of the human species alone, the subject must reinstate the phanic inner self's license to dispassionately and thoroughly reconsider all aspects of our commitment to phagic pseudo-worlds, beginning with questioning theoretical reason's foundational claim to be equipped to ontologize a single Nature-nature, a real-ideal infinite absolute, in and "as" necessity alone. We can also then reconsider how to engage with additional metarational questions, including the question of whether, why, and in which cases the world as it is in itself must be insulated (both cognitively or from the point of view of reason; *and* physically) from theorized ontologies of necessity and the "worlds" they posit in its stead, and where, if anywhere, the reverse must be allowed to continue. Given the deep entrenchment of CM, even these simple beginnings need the Controversy framing to anchor them.

K. C. Snow, *Our Inevitable Third Spinoza Controversy*,
https://doi.org/10.1007/978-3-032-03045-0_5

It is worth noting, too, as we discuss what could be the first steps along the neo-Jacobian path out of this third Controversy configuration, that if, and once, the metarational faculty is slowly revivified, the subject does not lose or discard the scientific knowledge we have gained in mathematical physics, but rather radically re-contextualizes it, along with any other vital products of theory. When the scientific model of the universe is taken as an absolute ontology, we refuse to acknowledge the distinction between its types of certainty (and its types of fully, but self-referentially, defined entities and terms) and anything like knowing with "certainty" what "must be" in our universe at its total scale such that it actually exists in any aspect of its being, including how and why it displays or shows those regularities we observe as its "laws," "matter," "energy," and so forth. For the neo-Jacobian position, this refusal [to distinguish between theory and world] has clearly played its hand to the very limit of its ability, and lost; its day has come and gone, and cost us much of the world. Necessity as nature has fully expended itself conceptually, coming full circle to end in an infinite multiverse which undermines the very notion of necessity itself. Necessitarian regimes have revealed themselves as theoretical, quintessentially human constructs; human ways of reasoning via mapping necessity relations (and generating necessity directives for the subject). When they are deployed in or as our actual "worlds", they have significant impacts, both on human cognition and decision-making, and on our environmental behaviors. The neo-Jacobian position does then insist that natural science consider through fresh eyes its claim to be able to ontologize and encompass the actual, historical universe within various rationalized and intelligible sets of laws and other necessity relations held to have "made" it or to "hold" it together. Still, if the Jacobian relocates reality and being back in the external world, before and beyond the grasp of theory, and recasts cosmology as physics and astronomy at large scales once more, this does not end these sciences' claims to be able to touch or reveal small aspects of reality, but instead bolsters their claims to realism by placing them back in touch with a reality allowed once more to be truly external to thought and free of necessitarian monism's deep layers of self-reference. The sciences only lose things such as their incoherent claim to know how to "make" reality in its entirety, or their overloading of their theoretical and observational findings with CM *a priori* assumptions about nature as ultimately a Spinozist Nature. The sciences, especially cosmology (and physics), are permitted, after two centuries, to put down the romantic monist burden they have never properly speaking been able to support.

Only in re-affirming the difference in kind between science and the universe itself can the metarational part of the subject find scope to return, since it is defined through contact with the real external, the Thou of the world which the I encounters before and beyond the closed circuits of any theoretical or conceptual analysis. As long as the universe remains replaced by the rationalized necessity-web of mathematical physics, as though this web were the same thing as the really existent universe, the metarational subject cannot find its bearings, even its reality; it remains dormant in the person, or begins to disappear altogether. Only once the metarational faculty is awake and present once more in the Western subject can the subject undertake a serious moral appraisal of the Western annihilation of the living world, since the metarational subject is also the seat of ethics and values. Only in reacknowledging the distance between science and something like "the whole", which is potentially really infinite (and not a mathematically infinite perfect generic), can the subject come to [re]cognize why and how theoretical ontologies do not actually allow it to work along anything like the inner seams of reality, and that the only standpoint of production it occupies can only ever be the standpoint of the human production of new forms, interventions, technologies, or [domesticated, synthetic, or informatic] spaces to the best of the theoretical understanding's ability. Only by acknowledging that science does not speak of being as such, and that the set of all possibles is neither a concept with any real existence nor an entity or set which bestows real existence on anything else through the relations of necessity it contains and generates, can the metarational subject regain the ability to restrict science to finite aims and enjoin the theoretical reason to cease vanishing, both cognitively and physically, the real external physical world.

The Controversy framing allows us to see that the annihilation of the physical world around us is inextricably part of taking the idea of "necessity" sitting at the root of a totalized, divinized, self-grounding ontology that Spinoza and others had explored in a theological context, and transmuting it into romantic philosophy of nature and thence into physics and cosmology. In then asserting, incoherently, that the resulting ontology of necessity accounted for all of being, requiring an end of the phanic, metarational capacity that might easily judge such an assertion fictitious, scientific "ontologizing" laid down the first precondition of the destruction of the actual world around us. By ending the phanic, metarational capacity to see the application of scientific necessity to being or the universe itself as erroneous, scientific reason was increasingly free to destroy the actual world around us to force it to match and uphold our regimes of necessity. Once theory had destroyed its [metarational] Other within the subject, there was no part of the subject which could truly contest this imposition of necessitarian certainty-regimes on the earth, just as there was little if any active resistance left to their dominance within the subject's full person, no "whole inner self" left to "recoil" from them, in Jacobi's phrasing (cf. Chs. I and II). Re-establishing the part of the person capable of establishing a dynamic back and forth exchange with the external world as a Thou, and of delimiting the automatic boundaries inherent in any schema built on necessity-relations – delimiting rationally what necessity-relations can and cannot signify – for a Jacobian constitutes the only viable first step in forming a true counter-possibility to the continued annihilation of the external and its replacement with the internal.

What a neo-Jacobian position could then potentially unlock, for our understanding and preservation of the non-human world, appears to be significant. The Jacobian is able to take up the stance that human beings inhabit a really infinite natural world whose actual being and ultimate manner of existing is completely unknown to us, no matter if we have figured out some sets of natural laws which make it easy for us to intervene in the world and test some few filaments of its infinite depth in certain useful ways. The Jacobian position would permit re-establishing a hierarchy of metarational control over the necessity-mapping faculty in the person, and this could, again, be expected to change the ways in which Western subjects intervene in the real world, including setting up spatial correlates (like disenclosure or rewilding) of subjects rediscovering new-old freedoms from systems of necessity which for over two centuries have gradually removed the sustained contact between the person and the real. Since believing like a CM naturalist in the infinite or absolute status of the "complex that we call *nature*," in Jacobi's phrase,[1] can only ever (for a Jacobian) annihilate what really is (cf. Chaps. 1–2), the Jacobian requires a non-philosophy of non-nature, which includes within it appreciation for the discoveries of science and understands them, but does not mistake them for explanations of being as such. The world as non-nature regains its independence from us, and we have to reconsider our manner of approaching and understanding it. Beginning from a position of acknowledgment of our radical ignorance in a Barthian and a Jacobian sense, we can end up with a different view of both non-nature and the human, and

[1] Jacobi, Concerning the Doctrine, 376 (no. 425).

organize our societies and our time differently. Allowing more real non-nature to physically return all around us, and then preserving it, could be anticipated to result, since the metarational subject requires, non-negotiably and as its pre-grounding, a Thou – which we have begun to lay out in this work is potentially inclusive of the physical world beyond the categories of theory, just as for Jacobi it entailed God (and other people). Only if we preserve the living (and nonliving) layer of the truly external world, the world human beings have not made, from rampant human conversion and vanishing, can our fuller selves even have a chance to encounter the world as a Thou, and as cognitively and physically a part of the primal and ineluctable meaning and, in a sense, method, of our existence as beings. If the subject can thus *directly* experience the external as the source of something crucial for the continuation of its (newly regained) fuller subjecthood and life, the fragile uncertainty of our odds of survival in a world we seem to be using all the tools and directives of our ontologies of necessity to actively dismantle will effortlessly and immediately come clearly to our mind's eye.

Part of the irony of the contemporary devastation and removal of the real world is that techno-scientific reason also prides itself on understanding the laws or conditions on earth which dictate whether and how we can survive. Portions of the scientific community may, including through legacy survival of the metarational capacity; but for the most part, what the scientific-technological view understands how to assess is not real survival but *production* and *necessity* according to the laws which inhere in the "complex that we call *nature*" (in Jacobi's phrase, again) and in the human-built layers we have inserted (by force) into the real world in accordance with building out this "complex" to our ostensible advantage.

The situation we are in now, environmentally, calls out for environmental cognition of a metarational type perhaps even more than at any time in the past, while also facing greater obstacles to just this type of cognition than it has ever faced. We have never in the past faced the *accumulated* effects of two (or five, or more, depending on where one begins the count) centuries of physical nihilism toward the living world, with the last century in particular entailing such an unprecedented expansion in scale, tempo, and technological interfaces' reach alongside all the other combined effects of so many humans being alive at once, consuming so much. Jacobi's (and Barth's) contribution to helping us more fully understand the crisis we have made for ourselves helps us at the very outset to press against the dominant assumption that we have most things right, but only have to make some minor adjustments to our way of inhabiting the earth in order to continue to "progress." Our failure, our forgetting, of environmental cognition is our most important failure to date; it is more important to understand this failure, even at the cost of overturning long-held assumptions and hubristic "nearly-there" illusions, than it is to make sure that our contemporary "worlds" can (somehow, impossibly) continue for just a while longer undisturbed by the external real.

The Spinoza Controversy presents as a crisis which begins in a fairly simple choice: whether or not to continue to believe in our own internalist version of nature as actually implicative of real being, of the real universe around us. The Jacobian perspective helps us understand that given how the West has twice chosen in both its prior Controversy iterations, it is not a surprise, and it is not only a minor error needing only a small course correction within a continued adherence to and real belief in CM, that we have ended up with a rapidly disappearing external world; nor is it a surprise or something which only needs a minor adjustment that the human subject is not just caught within three different kinds of internalist, unreal, but necessitarian and so enforceable, "worlds," but also already experiencing, as a result of this, a severe loss of individual freedom and subjecthood, as well as facing the future prospect of the species losing existence and life itself – whether the loss of health and life due to the contamination of our world, or the overall loss of life at the species level which the threat of the extinction of our own and so many other species represents (cf. Chap. 4). It is also not surprising, nor is it simple to correct, that given the triumvirate of being denied the real external natural world beyond us; being made subjects of the economic value-raising necessity-regime; and being made to at least witness and navigate, if not ourselves also resort to, the proliferation of private rationalities, now true, raw chaos, not pre-formal and not obeisant to the CM worldview, is beginning to characterize the experience of the Western subject, threatening to undermine the whole contemporary social order, as theoretical reason's "worlds" reveal their actual unreality, their ill-suitedness to life and to survival, a fact ever more poorly concealed by the day beneath their regimes of violent fabrication and enforcement. Without the metarational subject continually deciding on the use versus nonuse of, and setting limits on the interpretive license taken by, the theoretical understanding, necessity regimes have become not sense-making of the world but nonsense-making.

The neo-Jacobian path, again, requires a departure from the CM understanding and use of necessity relations and the imagined "principle" of necessity [as inhering in an infinite Nature]. Part of moving out of the Controversy along this new path is coming to see necessity statements like any other logical statements, which may or may not be shown to be true according to the terms established, not as though they are possessed of raw ontological power in themselves. The pre-Controversy context in which necessity, and causal necessity, were assigned ontological heft was completely different and deeply theologically imbued, including in Spinoza's own case (cf. Chap. 2); the ontological connotation does not apply in naturalism (nor, arguably, did it even successfully work in the theological context), and no amount of neo-Spinozist framing can change this. From the neo-Jacobian position we can understand why ontologies of necessity are always inherently nihilistic vis-à-vis the real environment beyond us, and seek different foundations for environmental cognition based in the fuller person, not in the correctness of any theory or its implicit framework. Nature as a whole or as such is not the type of thing that a theory can ever be correct about. We have to expand our use of reason to be able to see this, rather than taking the path of least resistance which assumes in monist fashion that all types of ideas are commensurable, and all facets of the subject are reducible to the logical faculty and its highly straitened version of thought. Jacobi's argument is that our use of reason can and must be far more expansive than it is permitted to be in CM naturalism, not less. He does not hold up the idea of a non-philosophy of non-knowledge because he is a theist; he holds up the template or aspiration for this new type of non-philosophy because he wants to distinguish this philosophy as a clearing away of the theoretical pseudo-world to enable us to encounter the real one again; he wants to distinguish his position as supra-theoretical so that he can consider himself an authentic realist and externalist, but also a restorer of the fuller subject and its more complete view of, and ability to encounter, reality. The non-human and unintelligible is what we must point to as reality, and thus is what we must first and foremost be realist "about." Once its raw externality is restored, the further CM error of perceiving necessity relations as generative will also automatically fall away, and being and generative power as such can be reconsidered on their own terms, and as really inhering in the universe itself, apart from the human; reconsidered, also, without recourse to any CM claims to be able to theorize or model them in terms of logical necessity or physical force.

If in the original Controversy, Spinozism was opposed mostly in philosophical terms, and in the second, in theological ones, the iteration has now arrived in which it must somehow be opposed in the external world's terms, the terms of the site where the consequences of the choice the West makes this time will physically play out, just as they have done following both previous Controversy iterations [to such enormously ill effect]. The Controversy "configuration" (to use Lazier's term) of today does not need the neo-Jacobian side, however, to put forward any new system of rationalizable, programmatic "environmental ethics", but only to show how the subject as such, and the entirety of Western reason itself, will not and cannot survive without *greatly increased freedom from, and inherent limiting of the presumed reach and significance of, necessitarian pseudo-worlds.* What the Controversy arguably surfaces this third and last time around is that the Western subject should either seek to reinstate its meta-rational faculties, to enable Jacobian environmental cognition as a form of spiritual cognition, and to be able to once more understand the limits inherent in the "worlds" created by theory, or else expect to continue to face the spatial, environmental, and social consequences of this form of cognition's continued absence.

The Spinoza Controversy is, I have been trying to argue in this work, a dispute over how one should understand the different types of reasoning at the subject's disposal as a person, not an academic debate over which overarching theory of nature or reality is best. Jacobi's position, twice defeated in the mainstream, has tended to survive best away from centers of power which seek to reduce all discourse to just these academic types of debates (so that the more powerful necessity-based argument can win without real thought being required, and necessity as sole standard for meaning and action will not itself be challenged), including both universities and seats of financial and political power, which have become increasingly (intellectually, physically and financially) invested since 1800 in the necessity-relations view of the real, the types of contained debates this view permits (which do not, again, seek to actually go against or dislodge necessity-regimes as such), and the tangible forms of power that these centers can gain by continuing to enthrone techno-theoretic human reason and dethrone the natural world and its reality. Jacobians tend to melt away from these centers, and derive meaning in existence from constant conversation with the external world. Now, as the last remaining inches of wilderness and last handfuls of wild animal habitats effectively face the bulldozer, and the West is rapidly becoming a spent and human-saturated world where nonsensicality and the end of Western reason itself increasingly threaten, the Jacobians in their rags must, as it were, return to the gleaming, uniform halls of the new type of *polis* to try one last time to present their case.

We are now at a different juncture from 1800 or 1920. Though each Controversy has in a sense been about how to regenerate and re-regulate reason, now this regeneration and re-regulation *must* somehow, as an ironclad prerequisite for either side claiming victory, also show it can incorporate regeneration of the external, non-human world as well, as part of the same regenerative process, and show it can do so *indisputably* and immediately, and not only ideally or potentially. The stakes this time around are not only cultural or spiritual, though they are this too; the stakes are the survival of the metarational subject, the survival of Western reason itself in both its theoretical *and* metarational forms, and the survival (and the "quality" of the survival) of much of the living (and some of the nonliving) layer of the earthly.

Ultimately the question whether the Controversy might be arbitrated differently this third time around turns on whether the Western subject still retains the capacity to question theory as ontology, or if the romanticization of theory has carried on for too long (and in an atmosphere almost completely deprived of any potential competing sources of meaning for human existence or for human society writ large). If the metarational balance is not restored to the subject and the CM absolutist conceits are not peeled away from nature, the romanticization of theory may still soon break down due to different, less positive and intentional pressures, as the West faces a further descent into an anarchy of autarkic rationalities, increasingly with no content and no meaning at all and cultivated ever more overtly and directly as fodder for the "Growth!" regime which has no human master, until, as Jappe warns, society fully devours itself.

A Neo-Jacobian Critique of Some Aspects of the Self-Interpretation of Big Bang Cosmology

Using a neo-Jacobian position to respond to big bang cosmology's use of relations of necessity to ontologize and speak about the generation of real things that exist, or claim to delineate things like the ostensible logical principles and generics attaching to infinite Nature, offers one potential starting-point for opening up new ground for environmental cognition. Since most of the outlines of a potential Jacobian opposition to interpreting any ontology of necessity (including big bang cosmology's) as actual and real have already been sketched in the foregoing chapters, I will only briefly recapitulate it here.

Jacobi's primary designation of the violence inherent in substituting our constructed world for the real one shows us the beginning, the first sign, of what big bang cosmology overlooks and fails to acknowledge. The choice of either fully committing to, or else rejecting, the act of rationally world-building (using theory and graspable necessity-relations alone) is by Jacobi's standard (cf. Chaps. 1–2) the most crucial free choice. If and when the subject decides to commit to the rationalized ontology of necessity as the real world, all truths available to the subject after that can only embody the rules of physics' "language-game", as later Wittgenstein might characterize it.[2] In an echo of Jacobi, Wittgenstein holds that the way "necessity" is used in any self-consistent, coherent logical system is rather anemic – it just means the kinds of "intrinsic necessity" we can only ever represent in language via "an arbitrary rule[3]" we make ourselves. This seems to echo Jacobi's much earlier resistance to the very notion that in mapping necessity we can resemble real nature as such. For Jacobi's part, being as such is not reachable even using Spinoza's own original approach; the existence of the real universe is only a truth we can encounter directly, and only as a fuller, metarational, person, one capable of interacting with it as a Thou. Being cannot be "met" in human-specified logical relations among various observed necessities and their re-statements; it cannot be met within the technological, synthetic world these control schemata have allowed us to build and impose on the rest of being – on the non-human realm and on the rest of the person. The back-and-forth between experiencing being, and reflecting on this experiencing of being, including reflecting on the distinction between thought of this kind (with its emphasis on limits and ignorance) and the necessitarian language-game of science, is the site where we could imagine something like Jacobian environmental cognition taking place.

The Jacobian position first and foremost asks for a distinction to be preserved between what is observed and what makes the world wherein the observation takes place. While we have no way of "recognizing anything as being, essentially, the outcome of an infinite process" or "recognizing any process as being infinitely old[4]", Moore means these statements

[2] Throughout *Philosophical Investigations* Wittgenstein discusses why the notion of a language-game is key for understanding our use of language. Cf., e.g., Ludwig Wittgenstein, 1958, *Philosophical Investigations,* trans. G. E. M. Anscombe, Englewood Cliffs: Prentice Hall, 5 (no. 7), or 167 (nos. 654–656).

[3] Wittgenstein, *Philosophical Investigations,* 116 (no. 372).

[4] A. W. Moore, 2001, *The Infinite,* London: Routledge, 216.

from the point of view of theoretical reason; there is nothing barring the metarational person from encountering the night sky as a really realized infinity which is, and is apart from any rules or necessity-patterns within it which observations might reveal. And the same metarational faculty can curtail claims by theoretical reason to attach to it as a single "unity" those generics or principles extrapolated from finite concepts (cf. Chap. 4).

Likewise, being as such, just like realized infinity as such, can remain a meaningful idea for the metarational faculty if we resist its reduction into a mere inherent property of necessity-relations or their imagined free-standing algorithmic unfolding "as" the world in the form of natural laws, and reject Wheeler's (and others', cf. Chap. 4) information view of the universe which would reduce being to information. The question of what "must" be is a closed-loop type and manufactured type of question [which is how we physically succeed in literally building out our manufactured goods and cities]; the question of what really is cannot be submitted to this imagined standard, for the neo-Jacobian position. The actually infinite is part of being, and like being, it does not merge with or even make sense to the theoretical sub-faculty.[5] We can only let it break in upon us, as Barth would say; we can only cognize it, or as Jacobi might say, recognize it. Cognition of the infinite, of being, of the external world as truly external and as truly existent, can be other than superimposition.

Cosmologists' surmisings about the universe we inhabit having potentially not existed, whether it was fine-tuned, or which additional infinitarian generalities or "principles" being tacked on to the model might make it sufficiently "necessary", belong to the type of theoretical speculation which to a Jacobian is impossible and meaningless; it falsely places the dictates of our ideal concept of necessity, or our conviction or intuition regarding how necessity operates in nature, above the reality of the universe itself, such that we could meaningfully speak about it not existing (in order to satisfy some requirement coming from the side of our theory, or our understanding of necessity). The set of all possibles is only an idealization and cannot be used to justify absolute counterfactual scenarios or thought experiments regarding the presence or absence of our world as a whole. For Jacobi, we can recall from Chap. 2, any talk of the "*possibility of the existence of nature*" impossibly seeks to entail absolute speech, or

[5] This is indirectly suggested by Aristotle, *Physics,* Book III, Part 5, where he concludes that there must be "no body which is actually infinite" since it leads to logical contradiction.

speech of the "unconditional[6]"; causal necessity applied in such a context becomes a monstrous imitation of its true self, and even less rational than a theistic account of how or why nature exists.[7] This is also reminiscent of Barth's stricture (per Chap. 3) on posing questions as to the origin of the "possibility of the very existence of any biological function[8]" using rational deduction. To the extent cosmology, in multiverse theory or even apart from it, thinks its necessity-webs can actually entail or presence these kinds of questions, it falls into categorical error and unreasonableness.

The general themes of a potential Jacobian critique of big bang cosmology are not wholly without precedent from other quarters. The philosopher Jacques Merleau-Ponty[9] observed in 1969 that using transcendental reasoning in the simplest Kantian sense is one of the ways in which logic is useful in building scientific cosmological systems, and that the very assertion that cosmology is *possible* (as a 'science') is a transcendental position, one which already "considerably limits the domain of these structures [spacetime structures admissible as the 'universe'] without completely determining them[10]". In 1994, Merleau-Ponty problematized making any straightforward comparison between scientific cosmology and traditional mythology, observing that while mythology may be empirically doubtful, it is not nonsensical, while scientific cosmology's very foundation is irrational because it is based in mathematical idealism; for Merleau-Ponty it is this "irrationality of its foundation and not any [empirical] unlikeliness of its hypotheses that makes cosmology unique[11]".

Stevenson compares scientific cosmology's belief that there could be a single explanatory theory of the universe to "theocratic transcendental

[6] Jacobi, Concerning the Doctrine, 376f.

[7] Jacobi, Edward Allwill's, 495.

[8] Barth, *Epistle,* 425.

[9] A cousin of the more famous philosopher Maurice Merleau-Ponty.

[10] Jacques Merleau-Ponty, 1969, Logique, Mathématiques Cosmologie, *Les Études philosophiques,* 4: 501–502. Translation mine.

[11] Jacques Merleau-Ponty, 1994, La cosmologie contemporaine doit-elle intéresser les philosophes?, *Revue Philosophique de la France et de l'Étranger,* 184: 290. Translation mine.

In general, and as briefly noted in Chap. 4, since the outset of big bang cosmology, there has always been a coterie of specialists on myth and/or history who are less than comfortable with any claims that the theory of the big bang offers a mythological account. This is consistent with the fact that big bang cosmology's myth-talk is akin to a new version of the German romantic new mythology of reason, as we have been discussing; such a project, as noted in the preceding chapters, is really about claiming to *end* myth in the "old" sense.

realism".[12] Życiński holds that big bang cosmology's use of the AP is idealistic and teleological.[13] Philosopher of science Fidelman notes that the Linde-ian infinite multiverse and the Everett interpretation-based many worlds theories of the cosmos are only Kantian ideas of pure reason, as evidenced by their entailment of paradoxes.[14] Philosopher of science Laino writes that loop quantum gravity, one candidate for a unified theory of everything, that would also function to describe a $t = 0$ scenario, seems to offer a "*reduction ad unum*" where "physical events and beings are only the apparent result of the intrinsic modifications" of a monistic and infinite whole,[15] suggesting not only the CM and early German romantic view, but even Jacobi's observation of necessity as always only expressing an identity relation effectively known and specified in advance, and never adding new knowledge or content (cf. Chap. 2, last section[16]). These perspectives suggest that other scholars have also at times found various elements of the CM positions of big bang cosmology problematic or puzzling, though without yet having the Jacobian frame of reference on hand as a possible means of expanding their consideration and critique of the model's romantic and necessitarian ontology.

CM's Olbers-Pangloss Universe

A simple example of the core of the Jacobian critique of big bang cosmology can be found in applying his perspective to an old question in night sky observation with which big bang cosmology claims continuity (e.g., continuing to propose solutions to the question using new data, and including a discussion of the case in standard cosmology textbooks[17]), but which dates from just after the first Controversy.

[12] Leslie Stevenson, 2015, A Theory of Everything? Kant speaks to Stephen Hawking, in *Inspirations from Kant: Essays*, ed. Leslie Stevenson, Oxford: Oxford University Press, 73, 74ff.

[13] Józef M. Życiński, 1987, The anthropic principle and teleological interpretations of nature, *Review of Metaphysics*, 41: 322f.

[14] Uri Fidelman, 2011, On the ontological status of some cosmological and physical theories, *The Journal of Mind and Behavior*, 32: 353–354, 357.

[15] Luigi Laino, 2018, Is knowledge of physical reality still Kantian? Some remarks about the transcendental character of loop quantum gravity, *Found Phys*, 48: 792–794.

[16] Regarding Jacobi wishing Kant had "kept steadfast in mind what scientific proofs can and cannot do," since Jacobi holds that by them "one can indeed elucidate cognition, but not increase it in content or ever secure a ground for it", Jacobi, On the Divine Things, 169.

[17] Cf., e.g., Jeffrey Bennett, et al., 2012, *The Essential Cosmic Perspective*, Boston: Addison Wesley, 492–493; or Liddle and Loveday, *Companion*, 222.

In the 1820s, a doctor and amateur astronomer named Wilhelm Olbers, influenced by the young Kant who had surmised an infinite universe (essentially for theistic reasons), and whom Olbers opens by citing approvingly, wrote a short essay in which he noted that while he assumed with Kant that in line with God's infinity, the real physical universe must also be infinite, to Olbers (unlike to Kant) the darkness of the night sky was not simple to explain if this were the case.[18] To Olbers, it seemed that an infinite collection of stars, no matter how sparsely scattered, would necessarily lead to a bright, not a dark, night sky.[19] This came to be known as Olbers' Paradox in later scientific cosmology. Olbers himself solved the question by inferring that it must be that "space is not… absolutely transparent," and this saves our night sky from being a "a shining ground of a solar brightness[20]", since the light from most stars loses its effective luster-for-us as it shines toward us through this partly opaque medium. Cosmologists today substitute the universe's finite past history for space's opacity, "solving" Olber's Paradox with a scientific explanation of why the universe "appears" finite (in terms of the night sky being mostly dark).

The actual paradox here, however, for a Jacobian, is epistemological and insoluble. Starting with open-ended and even theological

[18] Cf. Immanuel Kant, 2012, *Universal Natural History and Theory of the Heavens,* in *Kant: Natural Science: The Cambridge Edition of the Works of Immanuel Kant,* ed. E. Watkins, Cambridge: Cambridge University Press, 221 (I:255). In speculating (correctly) that distant "nebulous" stars were yet more galaxies like our own, Kant continues:

> … the weakness of the light … requires us to presuppose an infinite distance: all this is in agreement with holding the elliptical figures to be the same solar systems and, so to speak, Milky Ways, the constitution of which we have just developed ….

Now the attention of the observers of the heavens has enough motivation to occupy themselves with this suggestion. The fixed stars, as we know, all relate to a common plane and thus constitute an orderly whole, which is a world of worlds. One can see that in the immeasurable distances, there are more such star systems, and that creation in the entire infinite scope of its size is everywhere systematic and interrelated.

Though Kant famously would later, in *Critique of Pure Reason* (1781) effectively declare rational cosmologizing impossible (since neither an infinite nor a finite cosmos was fully graspable logically), in this earlier text (*Universal Natural History*) he was still at home with cosmologizing, albeit in a loose, genericizing way which also incorporated strong elements of theism.

[19] Cf. Heinrich W. M. Olbers, 1826, On the Transparency of Space by Dr. Olbers of Bremen, *The Edinburgh New Philosophical Journal*, 1: 141–150.

[20] Olbers, On the Transparency, 144.

considerations shared with Kant, and seemingly bolstered by aesthetic intuition, for example an intuition that "the beautiful order which we observe extending as far as our faculty of sight can penetrate, reigns equally through all space[21]", Olbers then needs a closed system to test his hypothesis (based on necessity-relations), and thus transitions over to *making* the universe behave like a closed and finite set of possibilities, despite, again, opening with the premise that it is not this type of a place or reality. He also first searches for reasons available to observation that would "induce us to abandon this opinion[22]" of the universe's infinity. This implies he wants to identify other things he might observe that would be evidence of finitude; though how he would confirm this is unclear. Indeed, the whole essay is devoted to taking the one piece of evidence he does at first have (in his assessment) of the night sky's apparent finitude – its darkness – and changing this into evidence of the night sky's infinitude, just as he had decided a priori the night sky must be. He accomplishes this reversal of his interpretation of the evidence by transmuting his question, whether the sky is dark because the night sky is finite, into a different, much more well-defined question, that of how quickly light might dissipate as it traveled through a slightly opaque medium. He hypothesizes that perhaps space is not transparent and thus dilutes the starlight, then tests the distance *from Earth* at which starlight would cease to be visible in such a medium. He chooses a known star with an estimated distance from Earth, Sirius, and calculates that if we took the combination of a distance 30,000 times greater than this from Earth, and his guess for how opaque the medium of space might be, then at and beyond this limit-point starlight would not reach us at a detectable level; its light would be reduced by close enough to 100 percent to escape our eye.[23] Again, this is not a measure of the behavior of infinity, but a basic thought experiment along the lines of holding a light in a dark space and measuring how far the light has to be removed from the observer before it is no longer visible.

This type of approach is a kind of simple and schematic foreshadowing of how scientific cosmology would address the problem of handling [apparent] observational evidence of "finitude" or finite behaviors in the universe, while the universe was also simultaneously assumed to be infinite. By claiming to handle an infinite limit-condition, while not actually

[21] Olbers, On the Transparency, 143.
[22] Olbers, On the Transparency, 143.
[23] Olbers, On the Transparency, 148–149.

"touching" the infinite quantity or presence at all (instead calculating out some finite problem from Earth's perspective), Olbers somehow positions the human knower as in possession of a "behavior" of infinity, with the same level of certitude as the reason can calculate dissipation of light in a semi-opaque medium. In fact, we do not know how to attach any trait or rule or behavior to the infinite whole of nature; we cannot make sense of how an infinite number of really present stars would act, due to the internalist, always finite, at most genericized, quality of all our science. Again, science will later introduce the notion of a finite past age of the universe, and the finite speed of light, to stand in for Olbers' semi-opacity of space, but the basic rudiments of the approach are the same; something must "make" the light diminish down to a "finite" observable amount because our scientific reason cannot handle the infinite (other than as a genericized version of our knowledge of the finite); science always *requires* us to inhabit a finite "envelope" as our largest calculable universe. We rule out the functional raw presence of a really realized historical infinity when we approach the task of rationalized cosmology itself, from which everything external or Other to theory must be removed. What engaged Jacobi was how little warrant there actually was to impose the way in which our theoretical reason works upon something which cannot be comprehended as a really existent and really external entity, or even properly encountered, by this faculty at all. We can recall the distinction Einstein drew between himself and Newton when he set about launching the new scientific cosmological enterprise in 1917: Einstein proceeded from a Machian premise that nature *had* to be causally connectable and monistic, and that "spatial conditions at infinity" or "infinite boundary conditions" were not intelligible enough and therefore had to be made more so (Chap. 3[24]) – arguably, had to be made even more fully [functionally] finite. Though clothed in vastly more sophisticated terms than Olbers used, the step is the same; turning from invoking infinity thematically to grappling with a finite problem mathematically, without acknowledging the extraordinary sleight of hand.

Even the simpler Olbers' Paradox premise still in use today that "in a postulated infinite universe with an infinite number of stars, it can be reasoned that the night sky ought to be as bright as the daytime Sun[25]", is for

[24] And cf. also Kerszberg, *Invented*, 363; and Smeenk, Einstein's role, 239ff.

[25] Cf. Zaki Harari, 2019, Analytical Resolution of the Dark Night Sky (Olbers') Paradox, *Astron. Nachr.*, 340: 511.

the Jacobian position unwarranted. We arguably do not know how to "postulate", or set up necessity-linked "must" or "ought" claims, in connection with a really external and actualized infinite physical space or grouping. Interestingly, the James Webb Space Telescope (JWST) and other new telescopes are placing empirical pressure[26] on whether the amount of light and radiation we can now actually detect in our "dark" night sky with these newly sensitive lenses still matches what the big bang model predicts we should see given the finite amount of time which has elapsed since $t = 0$[27] and the finite speed of light, and it is possible that additional observations will place more pressure on the model's self-interpretation from within, just as I am here raising potential ways of subjecting it to Jacobian pressure from without.

Another usefully schematic illustration of a potential Jacobian critique of the irrationality of any ontological or totalizing cosmological extension of necessity-relations into a would-be framing of really existing, historically present things comes in Voltaire's famous 18th-century novel *Candide*. Jacobi's impatience with romantic and Spinozistic faith in rational cosmologizing of the real is presaged in Voltaire's critique of Leibnizian efforts to make a rational necessity-scheme apply to the actual realm of existence. The problem for both Jacobi and Voltaire is, arguably, not one of theodicy, nor these systems' inability to "include" free will. The problem is the way such schemes silence the metarational perspective and force the person to try to live via logical concepts alone, risking losing our reason and reasonableness (of any kind) altogether. For both Jacobi and Voltaire, humans lose both their freedom and their reason when they commit to trying to really live inside a system of logical necessity. It does not matter whether this system itself is determinist and fatalist or more

[26] Alongside other more theoretical and even some philosophical pressures being occasionally put on the model, including over the problem of having to posit a universe which is mostly dark matter and dark energy, the latter of which particularly sometimes seems to function largely as a placeholder for an unknown parameter which will make the model calculate properly in future, and even in some ways as a mere re-statement of the assertion that rationalized cosmology must be more possible than it thus far appears, and thus in future with new observations it is assumed it will be so. Cf. Bennett, *Cosmic Perspective,* 445ff.; Kroupa, Dark Matter Crisis; and Graeme Rhook and Mark Zangari, 1994, Should we believe in the Big Bang?: A critique of the integrity of modern cosmology, *PSA: Proceedings of the Biennial Meeting of the Philosophy of Science Association* 1: 228–237.

[27] Rajendra P. Gupta, 2023, *JWST* early Universe observations and ΛCDM cosmology, *Monthly Notices of the Royal Astronomical Society*, 524: 3385–3395; Harari, Analytical Resolution.

rewarding to free will (in some way); the falseness of the entire system when offered as an account of real existence is the problem; the human being who chooses to live within it, can only ever be left helpless to delimit or control the algorithm-like rhythm of its constant flow of directives or claims that somehow, the real, existing world, and the rationally-derived theory of that world, are one and the same thing. These claims render life something senseless (in effect, mere information). The self loses its dignity as a full person, as the metarational self – the seat of character and moral judgment as well as of the fuller sense of reason – is forced to capitulate to the algorithmic, automated logic the theoretical reason believes it has uncovered as really expressing the main principles of the world in which the subject finds itself.

First, the young Candide learns from his tutor, Dr. Pangloss, the Leibnizian cosmology:

> Master Pangloss taught the metaphysico-theo-logo-cosmolo-nigology. He could prove to admiration that there is no effect without a cause; and, that in this best of all possible worlds, the baron's castle was the most magnificent of all castles, and my lady the best of all possible baronesses. It is demonstrable, said he, that things cannot be otherwise than as they are; for as all things have been created for some end, they must necessarily be created for the best end. Observe, for instance, the nose is formed for spectacles, therefore we wear spectacles. The legs are visibly designed for stockings, accordingly we wear stockings. Stones were made to be hewn, and to construct castles, therefore My Lord has a magnificent castle ....[28]

Candide then grows up and begins to encounter the real world, and the necessitarian systemization of the real continues to be exposed as absurdly mismatched to the realm of experience.

> NEVER was anything so gallant, so well accoutred, so brilliant, and so finely disposed as the two armies. ... The musket bullets swept away, out of the best of all possible worlds, nine or ten thousand scoundrels that infested its surface. The bayonet was next the sufficient reason of the deaths of several thousands.[29]

[28] Voltaire, 1901, *Candide; or, The Optimist,* trans. William F. Fleming, in *The Works of Voltaire: A contemporary version, Vol. I,* Paris: E. R. DuMont, 62.

[29] Voltaire, *Candide,* 68.

From the Jacobian perspective, any scientific cosmology based on asserting necessity-relations as the real world ends up with absurdities readily identifiable as such to the metarational self, whether they resemble the indisputable generic refrains to which Candide must resort, like "everything is necessary in the best of all possible worlds,"[30] or the harder-edged "anything that can happen will happen an infinite number of times"[31] which arises as part of the required mental frame for any who affirm the infinite multiverse. The senselessness of both types of "world"-creating statement does not make them "Copernican" or bold, the bearers of particularly difficult but still *true* tidings of an opaque world. They are only proof of our irrationally absolutized belief in our own internalist "complex that we call *nature*" (again in Jacobi's phrase), belief which leads to us becoming trapped within hollow regimes of our own making, instead of inhabiting the world as it really is and really offers us an existence which defies this type of reductive, necessity-driven account.

Toward a Jacobian Critique of the Destruction of the Other

In 1953, German physicist Werner Heisenberg remarked, as though it were not regrettable but simply a neutral fact, that humans had come to the point in their history that they had so thoroughly altered the world that rather than nature, "we 'meet' no one but ourselves[32]"; in an even more actively approving tone, Heisenberg also noted that this was mirrored in science, where the focus of scientific progress was becoming less "nature" and more "our knowledge of nature[33]" itself, given the new press forward into understanding quantum mechanics. He was presenting both anthropocentered realities as inevitable signs of our progress as a civilization.

For a Jacobian, a world where the human meets only itself points back to the experience of attempting to really inhabit a *nihil* of reason, which taken as a place for being, for existing, can only ever be a non-world, a

[30] Voltaire, *Candide,* 210.

[31] Guth et al., Inflationary Paradigm, 115.

[32] Werner Heisenberg, 1964, Das Naturbild der heutigen Physik (The Idea of Nature in Contemporary Physics), Lecture to meeting of Bayerische Akademie der Schönen Künste, Munich, 17 Nov 1953, reprinted in *W Universitas,* 7: 282.

[33] Heisenberg, Das Naturbild, 283.

value-less and unreal world. If the Other, the Thou, cannot easily be found in experience, this deprives the human subject of the most important element of meaning in our existence, and denies our reason experience of the grounding conditions it needs in order to function with the full scope and elasticity of its life and powers, the full scope of its various forms of sight. As we have already made clear, Jacobi would consider it a grim type of proof of the arrival of an age of nihilism linked with the end of the metarational subject and the elevation of the "weaving of weaving" view, that humans have literally begun to cause the "annihilation of nature"[34] spatially and physically. Environmental cognition, with the whole person, is a limiting exercise first and foremost. Relations of necessity must be limited to their proper sphere, not elevated. The metarational faculty must also limit the perception of our omnipotence and our commitment to self-worship, and resume awareness of the Thou of the actual external world, not necessarily in a theistic sense, but in the sense of the raw prior-ness of the existing infinite non-nature in which we find ourselves. The non-nature which is all that is non-human must be preserved both by and for the metarational subject.

The metarational part of the person is also in a sense, as noted earlier, the wild part; it is the part which most adamantly refuses to *enter into* the necessity-system it knows the end of in advance, instead seeking freedom from its yoke; it is the most individual aspect of the person. The metarational, wild subject rejects the non-world from which Otherness is banished, because it would therein lose the source and framing of its own real existence. The requirement for some Otherness to remain on the earth has a correlate with extracting and using less; and with the preservation of wilderness as a zone free from the man-built, and thus a zone which preserves our ability – an ability our metarational selves, in contradistinction to Heisenberg, can, again, recognize as vital for the proper functioning of our reason – to "meet" more than only ourselves. In wilderness, human conceptual dominance does not apply, symbolizing the freedom to encounter the Other unencumbered by imagined networks of necessity which would falsely precondition the exchange. Wilderness can also be seen in the Barth-ian sense as the place where the human's tangent-point connection with the infinite might be sought or felt, a space of freedom (or even escape) from the Barthian "Death" and "Night" which humans

[34] Cf. G. Ceballos, A. H. Ehrlich, and P. R. Ehrlich, 2015, *The Annihilation of Nature: Human extinction of birds and mammals*, Baltimore: Johns Hopkins University Press.

have actually enacted on earth via our destruction of so much Otherness. Wilderness enables "Life" and "Day" because it allows the person to encounter and understand its own being and also from this vantage understand more directly the limits on theoretical reason's circumscription of reality, re-opening the subject's phanic eye onto the real. This includes seeing ourselves through the eyes of all the other forms of life on earth, for whom we are unknowable Thous. The inert, definitional, internalist character of the "weaving of weaving" view becomes pronounced and blatant.

For a neo-Jacobian, we recall, to be properly a realist means preserving one's connection to the real world as it is without us and apart from us, and this is both symbolized by, and carried directly into tangible view in, wilderness. Wilderness can further virtue and a love of the good, and found a moral life up to and including an ability to have a genuine relationship with truths that actually derive from outside the I; for Jacobi, all these vital aspects of life must come from experience of that which cannot be proven or theoretically reconstructed, but only recognized and encountered.

By contrast with a world inclusive of the Other, the world of Heisenberg's where we meet only ourselves is full of isolation of the person. Without being permitted communion with the Other, the fuller person collapses or becomes dormant, and the necessity-mapping theoretical causal reason – for the very reason that it is self-contained and *inhumane*, and so does not require the metarational subject in order to function – is all that remains of reason and the subject as a whole. And then, as the pseudo-worlds "made" of necessity-relations which are even less constrained than science in their application and methods – those worlds discussed above of the "Growth!" regime and the other, more subjective, necessitarian regimes designated on smaller scales by individuals – proliferate, there are no metarational subjects left who are able to see and understand and navigate the pseudo-worlds of others, further accelerating the isolation and even the breakdown of the Western subject. Only the metarational subject can communicate with other subjects truly; trained by its experience of the world as first and foremost an external Thou, it is capable of accepting and weighing incommensurables without falsely dissolving or resolving their distinctions. Without the metarational bridge, while one "world" might either mimic or reject the outcomes of another "world", neither will really be affected, because the metarational interactions and exchanges seated in the person will not occur; each world will increasingly wall itself off from all others, unless it is strong enough, like the economic necessity regime, to consume multiple subjective worlds within its own by force. In such a state of affairs, the limits on theoretical reason proposed by the metarational subject cannot be implemented, since the metarational subject does not properly or fully exist, and non-nature, the really external world, will continue to find itself subjected to ceaseless conversion and devouring by the increasingly fragmented human realm.

For Barth it was God that was Other to, and thus must "break in" upon, misguided theories of reality as such. Any efforts [by theoretical reason] to try to replace God with some "high, exalted force, known or knowable"[35] or to put stock in a "super-world... also... known and accessible to us",[36] for Barth dealt only in fictions. Jacobi shows us that it is also the world, the really external world, which can perform the same unmasking or exposure of the mismatch between ultimate reality or all of being, and theory, provided we approach the world with a metarational subjectivity. The restoration of non-nature around us which might result from loosing the person from the prison of defining existence solely via the three kinds of ontologies of necessity could actually approach the sense of a miracle in a non-theistic sense. So much "being" in excess of necessity-relations does the external world in fact contain, that, so long deprived of it, so long accustomed to trying to believe they can and must inhabit only the worlds of their own making and "meet" no one but themselves, the human being could be overwhelmed by the world's being in ways which resemble the miraculous, affecting their approach to the environment, again, not consciously, or as the result of a rationalized directive, but through the effects over time of sustained metarational environmental cognition, and even perhaps the arrival of some uncontained love for the world and all the species in it.

In Chap. 2 we discussed how the romantics sought a vocational, striving understanding of the infinite that could foster the urge to undertake enough activity, work, or other intervention, to *make* the ideas of logic and the "world" they helped the subject weave more real than the real world that is not conceptualizable but only able to be encountered. The romantics still tied this striving to art and contemplation, and to science and mathematics in their purer form. In the two Controversy configurations since, striving has been replaced by merely "doing"; following the dictates of necessity, within the systems which conceive of the world as

[35] Barth, *Epistle,* 36.
[36] Barth, *Epistle,* 44.

solely necessity. In the twenty-first century, the scientific form of Spinozistic naturalism, though dominant still, is showing signs of crumbling from within in the face of the co-option of the ontology of necessity which it asserted really made nature, by the system of capital and value-creation which Jappe describes, and the informatic and sheerly subjective "world" proliferation this value system encourages, in part for reasons of mining human lives themselves for value-raising and thus profit. The technological augmentation of making our internalist worlds feel real has led to a condition in which the self has to world-make ever more rapidly and determinedly to avoid drowning amidst a vast sea of empty fabricated worlds of others, worlds which it increasingly has no way of joining or even understanding.

It may be that only a turn toward Jacobian non-philosophy of non-nature could potentially allow the West to avoid this building nihilism and the anarchic future we are perilously close to entering. Overspending our causal reason, insisting on necessity [as our theoretical faculty alone can parse "necessity"] as the only and sufficient ontological component of actual being, philosophy and the sciences have now lost control of this ultimately quite portable and transmutable power of "world-building". Perhaps only by returning to the radical either/or, the unfamiliar terms, of the Controversy, can we contest the West's increasing fragmentation into merely an infinite multiverse of anarchic rationalities and pseudo-worlds alongside an economic regime.

It appears this iteration of the Controversy must – can only be – our last. And arguably, at this juncture, only a non-philosophy of non-nature that refuses to condemn us to being the merest information processers within an absolutized necessitarian monist universe, where we are not even reasoning through complete causal chains, but only minute fragments of them or less-than-infinitesimal variations on given premises; refuses to condemn the living world to being merely fuel to quite literally run this futile exercise into the ground, has any hope of traction radical enough to make any difference, or to extricate us from the Controversy in a lasting way. No argument that relies on relations of necessity, and tries to pit one version of a necessity-web against another, to argue that one rationalized system is a better or more "policy-relevant" one than another, can suffice. Necessity relations' worlds are not policy relevant because they are not real except as they pertain to our own technologies and interfaces, our past extraction and converson projects. What matters for whether we cease destroying the external world, is whether or not we choose to see it again, whether we rouse our metarational selves from sleep (or near-death) to wakefulness, whether we endeavor to balance the phagic with the phanic capacity in human reason once more.

We are humans with reasoning powers beyond the "weaving of weaving" view and the pseudo-challenge of learning how to splice every thread of the weave ever-finer. We are historical beings living through a period in which we are facing an actual risk, not a theorizable one, of a full loss of freedom and life for our own kind and all other complex species as a result of our own actions and inactions; our own misuse of reason and refusal to adjust it at a fundamental enough level. Returning to the Controversy, and choosing differently this time, may be the only alternative to being milled down into bits of data which bolster value, condemned to isolated pseudo-worlds devoid of Thous, and eventually, as the non-nature around us becomes overly taxed and degraded, potentially wiped off the evolutionary board by natural "forces" infinitely greater and yet infinitely subtler than those we sought to mimic and model as only the give and take of a few kinds of necessity-relations our theoretical capacities could perfectly isolate.

If we are to return to open-endedness in the biosphere, which is the same as saying, if we are to return to a thriving biosphere with its future still undetermined, a real biosphere, we must return our networks of necessity to where they belong, nested within our fuller reason, used in some cases and not others, and always in consultation with the metarational self about the appropriateness of the use to which they are about to be put. If we lose our entire Otherness-receiving, phanic capacity, if we lose all remaining awareness of the external, there will literally be no way for us to reason our way through the contemporary environmental crisis. Otherness even in progressive social theory has been rendered a negative instead of the high positive it is in Jacobi, perhaps reflecting the CM view's innate fear of considering the Other in this more important sense, the sense which should connect us in a blood oath *across distance* to non-nature as Other, that should root our urban, dead zone, civility in the spark of actually non-theorizable life. Once the blood oath which is connected to, which is one of the conditions of, the simple fact that we exist, returns to our awareness, we can also see that the value-raising regime is just as flimsy a tale as any Panglossian ourobouros; we can finally disassemble it, instead of continuing to allow ourselves and our societies and our environment to be disassembled by it.

If the physical infinite universe is what possesses the generative power that somehow over time gave us actual life, and thus in a sense is what stands ready to take it back from us if we continue to fail to environmentally cognize for much longer, the universe might even come to resemble the traditional "Spirit" that was one of the rightful Thous of whom the empowered metarational faculty should be instinctively and pre-analytically aware for Barth and Jacobi. We can take a cue from Barth's insistence that metarational effort does not yield comfort or answers, but discomfort and torment – "Religion ... cannot liberate us[37]" – to also set aside any strawman version of some minimizable "nature religion" when what would be necessary would be an actual, living disposition which breaks into anything systematized or dead, and is simply what is necessary for Thou-cognition and thus for properly grounded reason as a whole. Any real nature religion would not give any easy or caricatured version of peace to the subject or the thinking "I", just as it would eschew any version of the rationalized, science-generated "super-world" Barth faulted some for holding up in place of the divine. How can a human subject expect peace in a truly infinite universe? The only impetus is to brave seeing the world as it is, not to find peace; to bear witness to and actually encounter infinity so that it breaks in upon one. Deeper joy and delight may, depending on the individual, be mixed with deeper sorrow, but it raises the quality of freedom available in this life to re-learn to see non-nature in a way which, to quote Feyerabend, allows us to "find situations which... involve [our] entire being[38]". Barth says, "We cannot begin with Truth, for it is our beginning. We must therefore range ourselves under it and make up our minds to live with it: under its destructive assault and with its indestructible blessing[39]". Is there something similar available to us if we take the utterly prior and external infinite as our Truth? Is there a version of the destructive assault/indestructible blessing dialectical pairing of actual incommensurables – even a version of grace – which may arrive from the universe itself? For Barth, "Restless, and terribly shattering, grace completely overthrows the foundations of this world; and yet ... the operation of grace [is] no mere negation. Grace is altogether 'Yes'; it is salvation, comfort, and edification[40]". If we return to affirming the external, and

[37] Barth, *Epistle*, 276.

[38] Paul Feyerabend, 2018, *The Tyranny of Science*, Ed. Eric Oberheim, Cambridge: Polity Press, 13.

[39] Barth, *Epistle*, 288.

[40] Barth, *Epistle*, 103.

negating the merely internal, replacing it with the truly subjective, the world we have for so long substituted for the actual one, would be overthrown, destabilized. Yet something else would come in its stead that bestows an abundance of meaning, even as our universe of necessity would be shown to be meaningless as an ontology of what really exists and of how it really is present. If it is only the ending of the world that modernity has built on top of the real one that can, physically speaking, save us, is not our only choice whether we seek out and intentionally receive the return of some form of breaking-in by the external into the internal, or continue to pretend in a world "made" of necessity relations, our fictionalized algorithmic progression, is really what is, such that we must continue to mimic and envelop ourselves in its dead laws, laws which could never create or be life itself? Could not part of what we acquire by way of truth in this breaking-in be the opportunity and the need to genuinely re-learn to ask the question, "What, then, are we?", in Barth's phrase, as well as to reconsider how and why nature may be undeniably divine? Freedom from ontologies of necessity leads here, to a great and wide open plain of questions.

Perhaps the reason a real metaphysics, or really a true ante-physics, of the infinite physical world itself has not been permitted for many centuries is that a true ante-physics would not set up physics to "be" nature, nor would it enable the *pólis* to model its systems of control on an imagined natural order which was only one of necessity, only knowable by theoretical reason. It may be that the only way of actually upholding political order in a secular world where there is no divinity is to base it on certainty-circuits; while true open-ended recognition of nature as realized infinity must always "pierce" through these, in Barth's term. But as we have been discussing, certainties can only be controlled and maintained for a finite time; their solidity turns out eventually to be illusion, because identitarian reason eventually exposes itself as circular, and the natural world cannot really be a collection of knowable necessity relations. Matter, as Hemsterhuis said two centuries ago, is only a word (cf. Chap. 1), and if we press the exercise of ontologizing what is using closed loops of theory, we end up with the infinite multiverse where all is of the utmost necessity in the most necessary of all possible worlds, but necessity has thereby also lost its meaning, finally making it clear that *nothing* at all *is* in such a world. Certainties and necessities such as those crafted by theoretical reason are not stable world-foundations; they lack all actual force and being. Sooner or later, they reveal their actual arbitrariness and the violence to the meta-rational self and the external world they entail – even as they proliferate. In a world of full subjects, which had not for any monotheistic or other reasons moved divinity *outside* of the physical world we encounter every day, divinity would be everywhere. One might think that in that situation, the question of replacing the real world with one of our own internalist making, whether cognitively or physically, would not, and could never, arise.

The ultimate question for the fuller person who encounters the universe – what *all that is, in its is-ness,* its *being,* really is – if we avoid the nihilistic Panglosses of our own necessity-identity loops, is not easily answerable, but can still, as a question, ground and frame environmental cognition. It could be this question's very difficulty which has sent humans at varying points into contemplating ideal-real, perfectly necessary realms; but this turning-away is not, as it turns out, a neutral choice, but erroneous, and quite possibly also unsurvivable. In order to cease destroying our part of the real world, we have to return to difficult, even "impossible," open-ended questions. Through its very difficulty, environmental cognition regarding the infinite universe, and the humility this must instill in us, the awareness of our radical ignorance, can make applying realist environmental cognition to our tangible Earth-bound conditions perhaps seem easier by comparison, or at least feasible.

The divorce of reason from existence which Jacobi mourned, and warned of the consequences of, is now happening not in a philosophical, not in a theistic or soteriological, but in a plastics-in-our-daily-bread, teeming urban cities with no clear way for youth to participate in the value-raising regime or to find meaning in it even if they do, paving of the last scraps of coast and forest, dwindling of the vaquita porpoise down to its last handful of individuals since it dies in captivity, kind of way. If Hemsterhuis' mild warning, that the universe permitting us to find an order, does not mean this order permits us to find the universe, to "world-build" using such meager facts, had been heeded, perhaps we would not have arrived here. But the West dismissed this warning, twice. There is arguably only one more chance, this chance, to try to understand the warning and why it means we must resuscitate the phanic stand.

In his 1785 reconstruction of an exchange with Lessing on Spinoza which launched the original Controversy, Jacobi writes:

> I [i.e., Jacobi]: … even the greatest mind, if it wants to explain all things absolutely, to make them rhyme with each other according to distinct concepts and will not otherwise let anything stand, must run into absurdities.
>
> Lessing: And he who will not explain?

> I: He who does not want to explain what is incomprehensible, but only wants to know the boundary where it begins and just recognize that it is there— of such a one I believe that he gains the greatest room within himself for genuine human truth.
>
> Lessing: Words, dear Jacobi, words! The boundary that you want to establish does not allow of determination. And moreover, you give free play to phantasies, nonsense, obscurantism.
>
> I: I believe that that boundary can be defined. I have no intention of establishing a boundary, but only of finding one that is already established and leaving it in place. And as for nonsense, phantasies, obscurantism. …
>
> Lessing: These are to be found wherever confused concepts rule.
>
> I: And even more where fictitious concepts do. Even the blindest, most nonsensical faith, if not the stupidest, finds its high throne there. For once one has fallen in love with *certain* [emphasis added] explanations, one accepts blindly every consequence that can be drawn from an inference that one cannot invalidate— even if one must walk on one's head.[41]

Arguably our contemporary environmental condition resembles "walking on one's head" – trying to live as though we are only creatures of our own necessity-worlds and enframing *nihils*, and not beings in a real universe. Our addiction to "certain explanations", which Jacobi warned against, has begun to devour our sense-making, which requires a metarational context, the same context in which there can be intersubjective communication among subjects all appearing as Thous to one another; without this context, theory's worlds quickly lose all meaning. The world becomes "informatic," and knowledge and experience are diminished into information along with everything else.

When cosmologists created a universe imagined as only certainties, laws, and a Nature that was somehow only necessity (as they understood this) yet encompassed the supposed all, the supposed font of all reality, they arguably created what Jacobi would call the "fictitious concept" (cf. above) to end all fictitious concepts, a world embedded in an infinite nihil that also displayed a pattern for world-building using theory, the lines of a new naturalistic practice both total in its ambitions and claims to ontologize, yet still somehow replicable enough for any number of sub-"worlds" to arise and find amplification within its various applications. This practice places the human mind in a false position, not only because of the number of worlds it entails, each with their own increasingly clashing and

[41] Jacobi, Concerning the Doctrine, 194 (secs. 30–31).

inherently senseless necessity-rubrics, but even more fundamentally, because the subject has to count "as" the external world only things which do not resemble it – circular models of necessity-relations. As our reason shrinks to this size, and we lose the real external world with its Thous which speak only to our metarational person, the imitable-ness of causal reason's "worlds" accelerates; the more the external world is vanished and replaced by the synthetic and the necessity-derived, the easier it becomes to tell oneself that one lives only in a world where the person need only freely select an "inference that one cannot invalidate," and "choose every consequence that can be drawn" from it, to make one's own, autarkic world. These forms of secession, of course, will not deliver the subject anywhere except into an ever smaller prison, wherein eventually, the person will be reduced to a mere contributory entity to the informatic chaos on the horizon poised to further severely threaten the survival of reason itself.

In re-considering the Controversy from Jacobi's perspective, we can grasp it as truly the Nihilism Controversy, which we are now entering into the brunt of in its third and final – its completely external and environmental – form. If we do not acknowledge the Jacobian perspective that there is no path to ultimacy via the merely theoretical, if we keep believing that we need theory to locate the real rather than seeing that we only need the realm of direct experience, we will never crack the carapace of withered, yet chaotically multiplying, monist life. It is past time to realize that our ongoing destruction of the living environment on which we rely proves to us that – to take liberties with Bruno Latour's phrase (that we have never been modern[42]) – we have never been Copernican. Copernicanism is only a romantic aesthetic, a game we play to cast theoretical reason's achievements onto an imagined absolutized, CM plane; and all it has done vis-à-vis our survival prospects is to damage them, potentially catastrophically, by making environmental nihilism acted out against the external world seem ultimately part of the single inevitable true necessity principle which is. From the vantage of environmental cognition any of our regimes of necessity are necessary only by their own internal criteria. They can be punctured and broken open by the tangent-line of our own metarational judgment any time we allow them to be, and they can and will eventually be broken in to by the external world itself.

[42] Cf. Bruno Latour, 1993, We Have Never Been Modern; Cambridge, MA, Harvard University Press.

Environmental nihilism is not physically survivable in the long term, and that is perhaps how the external grace of the world affects the human being in the end. As environmental destruction worsens and the ontology of necessity proves ever more its fatal, rotten imitable-ness, there is less with each passing year to indicate that CM is the firm foundation upon which the future of the human world can rest. Novalis dreamed that abstraction and the rationalizing of the human being's relationship with the infinite would lift up our experience, writing "Before abstraction everything is one – but it is one as chaos is – after abstraction everything is again unified – but this unification is a free alliance of independent, self-determined beings. A crowd has become a society – chaos is transformed into a manifold world[43]". But the arc of this manifold world is as short as any dream of theory imposing itself on real life must be; as necessity regimes hollow out the world and enforce their own hollowness on the human person, there is no "society" forming, as Novalis dreamed, but instead a world of science-imitating separate rationality-worlds united only under the auspices of all adhering to *a* necessity regime – each to their own. Our society is disintegrating; human beings in the West, and in every society globally, ever more seem to constitute only a "crowd", not a "society", and it is a crowd which becomes each year ever more merely crowd-like, ever more chaotic – and not the romantic, pre-formal type of chaos, but the true, post-formal chaos of the sole data point, the chaos of the collapse of the meaningfulness of worlds as such.

The Controversy framing and the Jacobian alternative non-philosophy of non-nature – which can also speak of the infinity and the reality of the natural world, in ways which avoid seeking to systematize it or in any way make it resemble our causal loop forms of sub-reasoning – is currently noticeably lacking in today's discourse on the global environmental crisis. Even environmental studies does not teach environmental cognition, but only more of the same imagined relations of necessity as if they really "are" the [ailing] external world. In such necessity networks, anything can be altered by the human who intends to do so; the scientists planning possible planetary interventions resemble big bang cosmologists testing different possible universe models to see which best fits observations and which might lead to different outcomes when some parameter is changed.

It is worth considering whether the Spinoza Controversy's outcome both times to date has been the result of some kind of self-protective

[43] Novalis, Miscellaneous Observations, 40 (no. 94).

lashing-out against the metarational by a form of reasoning which proves more immediately useful as a way of ordering the world, and extracting energy and life from the realm of experience without moral or any other subjective hesitation. If this is the case, will the metarational alternative always be sidelined, eventually becoming a footnote to the onward expansion of the human juggernaut which has simplified its presence and way of living on earth by shrinking reason down to fit the new world of the only-human and, increasingly, the horizontalizing senselessness of the informatic? In a sense, this is the most chilling part of Jacobi's warning: the more we trust to theory, the more we lose the sense of why we should not do so. Already, today, it is only through a massive effort of *epektasis* that we can stretch toward understanding, feeling, the weight of what nature, too, like Barth's God, has to offer by way of recompense, once it is once more recognized as Other to our small knotted language of cause, necessity, and empty self-identity, and sought out as it is in itself, a seeking which for many or most Western subjects may feel like a process of tunneling out of an ice-cave through twenty feet of snow, to reach the surface at last. Triangulating a Jacobian non-philosophy or, we might say, a Jacobian rural philosophy, of non-nature, with Jacobi's original non-philosophy of non-knowledge, and Barth's theology of crisis, we have begun to lay out in this work the need for a return to the metarational view of the person, the earth, and the earth's actually infinite setting. The weaving of weaving view ends in a tangled, strangling, pile of thread, much like our tangled internet and what it pretends to generate. We have been subjected to the impossible task of trying to live in the world of annihilated reality which results when CM insists it can make worlds; we are drowning in synthetics of all types. Our cleverness at solving some problems is offset by all that we are losing. Neither the earth nor the subject can survive much longer under these internalist necessitarian regimes. All that is left is to start tunneling upwards.

Foregrounding the physical world as a real source of exteriority (in terms of its being) which cannot be conceptualized or made "necessary" in logic's sense via any imaginary "Copernican" re-positionings of the subject requires foregoing choosing to begin cosmology with *a priori* monism regarding what really *is*. It requires not building the neo-romantic cosmos as part of a single Nature-nature, and instead exploring ways in which we might reconsider the physical world immediately around us, within a larger acknowledgment of the potentially heterogeneous, and undefinable, character of the being which inheres in the universe. This being must be experienced by and as a human being, rather than speculated upon via theoretical reason, in order to be seen as it really *is*, including as actually and really infinite and yet specific, not in any way embodying any kind of generic case or claim. What this turn may require is discontinuing the romantic poetic-scientific-symbolic view focused on unity, and its replacement with a more circuitous and immersive – even divided and baroque – approach to the world we live in and morally should, as well as biologically must, share with millions of other species. Allowing the local world to remain baroque[44] and impenetrable, of course, also has an obvious spatial and ethical corollary of minimizing human removal of the ability of other "Thous" to survive, and ending the violent erection of fabricated worlds.

For Beiser, again, the underlying crux of the original Spinoza Controversy turned on the question "Why should we be loyal to reason if it pushes us into the abyss[45]?" The stakes then were at least on their face humanistic, society-based ones: [theoretical] reason was perceived as beginning to "fail" as a possible backbone for religion, the state, and morality (cf. Chap. 1)[46]. If the West can still re-arbitrate this question – if the Spinoza Controversy is still a live debate, as I have suggested here, and not a pre-scripted tragedy – we must also acknowledge that we now face an even deeper crisis of reason, and an even more absolute abyss. Jacobi's non-nature is arguably well on its way to arriving back at the doorstep; we will meet it either through a swift return to environmental cognition which leads us to see, and rapidly change, our environmental actions in time, or through its "breaking in" upon us in a different sense, when its ability to support our phagic reign simply by its own terms comes to a [likely highly "unpredictable" for us] close.

It seems the time has come to finally choose the twice-dismissed path out of the West's defining, recurring, Controversy, the path which can restore the metarational subject and resuscitate environmental cognition and the phanic stand. If the other choice is the *nihil*, we can only affirm the chimera.

[44] John Law, 2004, And if the global were small and noncoherent? Method, complexity, and the baroque, in *Environment and Planning D; Society and Space*, 22, 13–26.

[45] Frederick Beiser, 1993, *The Fate of Reason: German philosophy from Kant to Fichte,* Cambridge: Harvard University Press, 80.

[46] Beiser, German Idealism, 1.

References

Abrams, M. H. 1973. *Natural Supernaturalism: Tradition and Revolution in Romantic Literature.* New York: W. W. Norton.

Aguirre, Anthony. 2007. Making Predictions in a Multiverse: Conundrums, Dangers, Coincidences. In *Universe or Multiverse?* ed. Bernard Carr, 367–386. Cambridge: Cambridge University Press.

Aguirre, Anthony. 2011. Cosmological Intimations of Infinity. In *Infinity: New Research Frontiers*, ed. Michael Heller and W. Hugh Woodin, 176–192. Cambridge: Cambridge University Press.

Anzola, D., P. Barbrook-Johnson, and J. I. Cano. 2017. Self-Organization and Social Science. *Computational and Mathematical Organization Theory* 23:221–257.

Aristotle. 2006. *Physics.* Translated by R. P. Hardie and R. K. Gaye. http://classics.mit.edu/Aristotle/physics.html

Armstrong, Karen. 2005. *A Short History of Myth.* Edinburgh: Canongate.

Baark, Sigurd, and Christine Svinth-Vaerge Põder. 2023. Crisis and Reorientation: Introduction. In *Crisis and Reorientation: Karl Barth's Römerbrief in the Cultural and Intellectual Context of Post WWI Europe*, ed. Christine Svinth-Vaerge Põder and Sigurd Baark, 1–11. Cham: Palgrave Macmillan/Springer.

Baker, Jennifer. 2007. Natural Science and the Romanticisms. *ESQ: A Journal of the American Renaissance* 53:387–412.

Balashov, Yuri. 2009. A Cognizable Universe: Transcendental Arguments in Physical Cosmology. In *Constituting Objectivity: Transcendental Perspectives on Modern Physics*, ed. M. Bitbol, Pierre Kerszberg, and Jean Petitot, 269–278. Dordrecht: Springer Netherlands.

K. C. Snow, *Our Inevitable Third Spinoza Controversy*,
https://doi.org/10.1007/978-3-032-03045-0

Barnes, Luke. 2017. Testing the Multiverse: Bayes, Fine-Tuning and Typicality. In *The Philosophy of Cosmology*, ed. Khalil Chamcham, Joseph Silk, John Barrow, and Simon Saunders, 447–466. Cambridge: Cambridge University Press.

Barth, Karl. 1933. *The Epistle to the Romans.* Translated by Edwyn C. Hoskyns. Oxford: Oxford University Press.

Behler, Ernst. 1996. Friedrich Schlegel's Theory of an Alternating Principle prior to his Arrival in Jena (6 August 1796). *Revue Internationale de Philosophie* 50:383–402.

Beiser, Frederick. 1993. *The Fate of Reason: German Philosophy from Kant to Fichte.* Cambridge: Harvard University Press.

Beiser, Frederick. 2002. *German Idealism: The Struggle Against Subjectivism, 1781 – 1801.* Cambridge: Harvard University Press.

Beiser, Frederick. 2006. The Paradox of Romantic Metaphysics. In *Philosophical Romanticism*, ed. N. Kompridis, 217–237. London: Routledge.

Bell, David. 1984. *Spinoza in Germany from 1670 to the Age of Goethe.* London: University of London Institute of Germany Studies.

Belyaev, Alexey. 2017. A New Model of the Birth of the Universe. *Universal Journal of Physics and Application* 11:182–189.

Bennett, Jeffrey, Megan Donahue, Nicholas Schneider, and Mark Voit. 2012. *The Essential Cosmic Perspective.* Boston: Addison Wesley.

Blumenberg, Hans. 1985. *Work on Myth.* Translated by R. M. Wallace. Cambridge: MIT Press.

Borgmann, Albert. 2006. Broken Symmetries: The Romantic Search for a Moral Cosmology. In *Philosophical Romanticism*, ed. N. Kompridis, 238–262. London: Routledge.

Bowie, Andrew. 1993. *Schelling and Modern European Philosophy: An introduction.* London: Routledge.

Bowie, Andrew. 2000. The Romantic Connection: Neurath, the Frankfurt School, and Heidegger, Part I. *British Journal for the History of Philosophy* 8:275–298.

Bowie, Andrew. 2003. *Aesthetics and Subjectivity.* Manchester: Manchester University Press.

Bowie, Andrew. 2016. Friedrich Wilhelm Joseph von Schelling. In *The Stanford Encyclopedia of Philosophy*, ed. Edward N. Zalta. https://plato.stanford.edu/entries/schelling/

Breazeale, Dan. 2024. Johann Gottlieb Fichte. In *The Stanford Encyclopedia of Philosophy*, ed. Edward N. Zalta and Uri Nodelman. https://plato.stanford.edu/archives/fall2024/entries/johann-fichte/

Brierley, Michael. 2004. Naming a Quiet Revolution: The Panentheistic Turn in Modern Theology. In *In Whom We Live and Move and Have Our Being: Panentheistic Reflections on God's Presence in a Scientific World*, ed. P. Clayton and A. Peacocke, 1–15. Grand Rapids: William B. Eerdmans Publishing Company.

Bullock, Marcus Paul. 1980. *Eclipse of the Sun – Friedrich Schlegel, Walter Benjamin, from New Mythology to the Theory of New History*. PhD diss., University of Oregon. ProQuest.

Caneva, Kenneth L. 1997. Physics and 'Naturphilosophie': A Reconnaissance. *History of Science* 35:35–106.

Carr, Bernard. 2017. Black Holes, Cosmology and the Passage of Time. In *The Philosophy of Cosmology*, ed. Khalil Chamcham, Joseph Silk, John Barrow, and Simon Saunders, 40–65. Cambridge: Cambridge University Press.

Carroll, Sean. 2016. *The Big Picture: On the Origins of Life, Meaning and the Universe Itself*. London: Oneworld Publications.

Carroll, Sean. 2018. Purpose, Freedom, and the Laws of Nature. In *Neuroexistentialism: Meaning, Morals and Purpose in the Age of Neuroscience*, ed. G. Caruso and O. Flanagan, 299–308. Oxford: Oxford University Press.

Carroll, Sean. 2019. *Something Deeply Hidden: Quantum Worlds and the Emergence of Spacetime*. New York: Dutton.

Ceballos, Gerardo, Anne H. Ehrlich, and Paul R. Ehrlich. 2015. *The Annihilation of Nature: Human Extinction of Birds and Mammals*. Baltimore: Johns Hopkins University Press.

Ceballos, Gerardo, Paul R. Ehrlich, and Peter H. Raven. 2020. Vertebrates on the Brink as Indicators of Biological Annihilation and the Sixth Mass Extinction. *Proceedings of the National Academy of Sciences - PNAS* 117:13596–13602.

Chaisson, Eric. 2006. *Epic of Evolution: Seven Ages of the Cosmos*. New York: Columbia University Press.

Creath, Richard. 2017. Logical Empiricism. In *The Stanford Encyclopedia of Philosophy*, ed. Edward N. Zalta. https://plato.stanford.edu/entries/logical-empiricism/

Crowe, Benjamin D. 2008. On 'the Religion of the Visible Universe': Novalis and the Pantheism Controversy. *British Journal for the History of Philosophy* 16:125–146.

D'Espagnat, B. 1987. Empirical Reality, Empirical Causality, and the Measurement Problem. *Foundations of Physics* 17:507–529.

Damour, Thibault. 2005. Einstein 1905 – 1955: His Approach to Physics. In *Einstein, 1905-2005: Poincaré Seminar 2005*, ed. T. Damour, O. Darrigol, B. Duplantier, and V. Rivasseau, 151–182. Basel: Birkhäuser.

Davies, Paul. 2004. Teleology Without Teleology: Purpose Through Emergent Complexity. In *In Whom We Live and Move and Have our Being: Panentheistic Reflections on God's Presence in a Scientific World*, ed. P. Clayton and A. Peacocke, 95–108. Grand Rapids: William B. Eerdmans Publishing Company.

Davies, Paul. 2018. In Steve Paulson, Paul Davies, Ard Louis & Lucianne Walkowicz. A touch of awe: Crafting meaning from the wonder of the cosmos. *Annals of the New York Academy of Sciences* 1432(1):46–62.

Day, Aidan. 2012. *Romanticism: The New Critical Idiom*. London: Routledge.

Dicke, R. H. 1961. Dirac's Cosmology and Mach's Principle, Letter. *Nature* 192:440–441.

Dingle, Herbert. 1937. Modern Aristotelianism. *Nature* 139:784–786.

Dirac, P. A. M. 1938. A New Basis for Cosmology. *Proceedings of the Royal Society of London, Series A* 165:199–208.

Dirac, P. A. M. 2003. The Relation Between Mathematics and Physics, Address to the Royal Society of Edinburgh, February 6, 1939. Reprinted in *Resonance* 102–110.

Douglas, A. 2015. Was Spinoza A Naturalist? *Pacific Philosophical Quarterly* 96:77–99.

Duro, Paul. 2010. 'Great and Noble Ideas of the Moral Kind': Wright of Derby and the Scientific Sublime. *Art History* 33(4):660–679.

Eddington, A. S. 1920. The Meaning of Matter and the Laws of Nature According to the Theory of Relativity. *Mind* 29:145–158.

Eichner, Hans. 1982. The Rise of Modern Science and the Genesis of Romanticism. *PMLA* 97:8–30.

Einstein, Albert. 1952. *Cosmological Considerations in the General Theory of Relativity.* Translated by W. Perrett and G. B. Jeffery. Reprinted from *The Principle of Relativity*, ed. H. A. Lorentz, et al., 175–188. Dover. Available at http://einsteinpapers.press.princeton.edu, Vol. 6: The Berlin Years: Writings, 1914 – 1917 (English translation supplement), Doc. 43.

Einstein, Albert. 1982. Principles of Research. Reprinted from Albert Einstein, *Ideas and Opinions.* Translated by Sonja Bargmann, 224–227. New York: Crown. Available at http://einsteinpapers.press.princeton.edu, Vol 7: The Berlin Years: Writings, 1918-1921 (English translation supplement), Doc. 7.

Ellermann, Greg. 2015. Speculative Romanticism. *SubStance* 44:154–174.

Ellis, George F. R. 2017. The Domain of Cosmology and the Testing of Cosmological Theories. In *The Philosophy of Cosmology*, ed. Khalil Chamcham, Joseph Silk, John Barrow, and Simon Saunders, 3–39. Cambridge: Cambridge University Press.

Feldt, Laura. 2012. Wilderness in Mythology and Religion. In *Wilderness in Mythology and Religion: Approaching Religious Spatialities, Cosmologies, and Ideas of Wild Nature,* ed. Laura Feldt, 1–23. London: Equinox Pub.

Feyerabend, Paul. 2016. *Philosophy of Nature*, ed. Helmut Heit and Eric Oberheim. Translated by Dorothea Lotter and Andrew Cross. Cambridge: Polity Press.

Feyerabend, Paul. 2018. *The Tyranny of Science*, ed. Eric Oberheim. Cambridge: Polity Press.

Fidelman, Uri. 2011. On the Ontological Status of Some Cosmological and Physical Theories. *The Journal of Mind and Behavior* 32:351–359.

Fletcher, Michael-Shawn, Rebecca Hamilton, Wolfram Dressler, and Lisa Palmer. 2021. Indigenous knowledge and the shackles of wilderness. *PNAS* 118.

Förster, Eckart. 2012. Goethe's Spinozism. In *Spinoza and German Idealism*, ed. E. Förster and Y. Melamed, 85–99. Cambridge: Cambridge University Press.

Forster, Michael N. 2012. Herder and Spinoza. In *Spinoza and German Idealism*, ed. E. Förster and Y. Melamed, 59–84. Cambridge: Cambridge University Press.

Frank, Manfred. 2004. *The Philosophical Foundations of Early German Romanticism*. Translated by E. Millán-Zaibert. Albany: State University of New York Press.

Frank, Manfred. 2016. 'The Difficult Step into Actuality': On the Makings of an Early Romantic Realism. *Comparative and Continental Philosophy* 8:199–215.

Gale, George. 1975. Leibniz, Chew, and Wheeler on the Identity of Physical and Philosophical Inquiry. *Review of Metaphysics* 29:322–333.

Gare, Arran. 2011. From Kant to Schelling to Process Metaphysics: On the Way to Ecological Civilization. *Cosmos and History: The Journal of Natural and Social Philosophy* 7:26–69.

Gare, Arran. 2014. Deep Ecology, the Radical Enlightenment, and Ecological Civilization. *The Trumpeter* 30:184–205.

Gare, Arran. 2018. Natural Philosophy and the Sciences: Challenging Science's Tunnel Vision. *Philosophies* 3:87–115.

Garrard, Graeme. 2005. *Counter-Enlightenments: From the Eighteenth Century to the Present*. London: Routledge Taylor & Francis Group.

Giannetto, Enrico. 2009. The Electromagnetic Conception of Nature at the Root of the Special and General Relativity Theories and its Revolutionary Meaning. *Science & Education* 18:765–781.

Gleiser, Marcelo. 2005. *The Dancing Universe: From Creation Myths to the Big Bang*. Lebanon: Dartmouth College Press.

Gower, Barry. 1973. Speculation in Physics: The History and Practice of *Naturphilosophie*. *Studies in History and Philosophy of Science* 3:301–356.

Graham, Daniel W. [Trans. & Ed.]. 2010. The Texts of Early Greek Philosophy: The complete fragments and selected testimonies of the major Presocratics, Part I. Cambridge: Cambridge University Press.

Grant, Iain Hamilton. 2010. Introduction to Schelling's *On the World Soul*. In *Collapse* VI, ed. R. Mackay, 59–65. Falmouth: Urbanomic.

Greene, Brian. 2011. *The Hidden Reality: Parallel Universes and the Deep Laws of the Cosmos*. New York: Vintage.

Gupta, Rajendra P. 2023. *JWST* Early Universe Observations and ΛCDM Cosmology. *Monthly Notices of the Royal Astronomical Society* 524:3385–3395.

Guth, Alan, David Kaiser, and Yasunori Nomura. 2014. Inflationary Paradigm After Planck 2013. *Physics Letters B* 733:112–119.

Hadot, Pierre. 2006. *The Veil of Isis: An Essay on the History of the Idea of Nature*. Translated by Michael Chase. Cambridge: Belknap Press.

Hallyn, Fernand. 1990. *The Poetic Structure of the World: Copernicus and Kepler*. New York: Zone Books.

Halmi, Nicholas. 2007. *The Genealogy of the Romantic Symbol.* Oxford: Oxford University Press.

Halpern, Paul, and Nick Tomasello. 2016. Size of the Observable Universe. *Advances in Astrophysics* 1:135–137.

Hamann, Johann Georg. 2007. Metacritique on the Purism of Reason (written in 1784). In *Hamann: Writings on Philosophy and Language*, ed. K. Haynes, 205–218. Cambridge: Cambridge University Press.

Harari, Zaki. 2019. Analytical Resolution of the Dark Night Sky (Olbers') Paradox. *Astron. Nachr.* 340:510–524.

Hartle, James B. 1991. Excess Baggage. In *Elementary Particles and the Universe: Essays in Honor of Murray Gell-Mann*, ed. John Schwarz, 1–16. Cambridge: Cambridge University Press.

Heisenberg, W. 1964. Das Naturbild der heutigen Physik (The Idea of Nature in Contemporary Physics). Reprinted in *W Universitas* 7:273–287.

Hemsterhuis, François. 2022. Aristaeus, or on the Divinity. In *The Dialogues of François Hemsterhuis, 1778-1787*, ed. Jacob van Sluis and Daniel Whistler, 62–99. Edinburgh: Edinburgh University Press.

Hemsterhuis, François. 2023. *The Philosophical Correspondence and Unpublished Writings of François Hemsterhuis*, ed. Jacob van Sluis and Daniel Whistler. Edinburgh: Edinburgh University Press.

Herder, Johann Gottfried. 1940. *God, Some Conversations.* Translated by Frederick H. Burkhardt. Indianapolis: Bobbs-Merrill.

Herder, Johann Gottfried. 1966. *Outlines of a Philosophy of the History of Man* (Ideen zur Philosophie der Geschichte der Menschheit). Translated by T. Churchill. New York: Bergman Publishers.

Hinchman, Lewis P., and Sandra K. Hinchman. 2007. What we owe the Romantics. *Environmental Values* 16:333–354.

Hölderlin, Johann Christian Friedrich. 2009. *Friedrich Hölderlin: Essays and Letters*, ed. and Translated by Jeremy Adler and Charlie Louth. London: Penguin.

Hölderlin, Johann Christian Friedrich. 2019. *Hyperion, or the Hermit in Greece.* Translated by Howard Gaskill. Cambridge: Open Book Publishers.

Holton, Gerald. 2003. Einstein's Third Paradise. *Daedalus* 132:26–34.

Humboldt, Alexander. 1845 – 1858. *Kosmos. Entwurf einer physischen Weltbeschreibung.* Stuttgart und Tübingen: T. O. Gotta'scher Berlag.

Humboldt, Alexander. 1849–1858. *Cosmos: A Sketch of a Physical Description of the Universe.* Five Volumes. Translated by E. C. Otté. London: Henry G. Bohn.

Jacobi, Friedrich Heinrich. 2016. *The Main Philosophical Writings and the Novel* Allwill. Translated by George di Giovanni. Montreal & Kingston: McGill-Queen's University Press.

Jacobi, Friedrich Heinrich. 2024. *On the Divine Things and their Revelation.* Translated by Paolo Livieri. Montreal & Kingston: McGill-Queen's University Press.

Jaki, Stanley L. 1989. *God and the Cosmologists.* Edinburgh: Scottish Academic Press.

Jappe, Anselm. 2023. *The Self-Devouring Society: Capitalism, Narcissism, and Self-destruction*. Translated by Eric-John Russell. Brooklyn: Common Notions.

Jeans, James. 1932. The Mathematical Aspect of the Universe. *Philosophy* 7:3–14.

Josephson-Storm, Jason. 2018. *The Myth of Disenchantment: Magic, Modernity, and the Birth of the Human Sciences*. Chicago: University of Chicago Press.

Kant, Immanuel. 1881. *Critique of Pure Reason*. Translated by F. Max Müller. New York: Macmillan.

Kant, Immanuel. 2012. Universal Natural History and Theory of the Heavens (1755). In *Kant: Natural Science*, ed. E. Watkins, 182–308. Cambridge: Cambridge University Press.

Keedus, Liisi. 2023. 'As a Tangent Touching a Circle': Karl Barth and Dialectical Theologians Rethinking Time After 1918. In *Crisis and Reorientation: Karl Barth's Römerbrief in the Cultural and Intellectual Context of Post WWI Europe*, ed. Christine Svinth-Vaerge Põder and Sigurd Baark, 35–38. Cham: Palgrave Macmillan/Springer.

Kenney, John P. 2005. *The Mysticism of Saint Augustine: Rereading the Confessions*. New York: Routledge.

Kerszberg, Pierre. 1989. *The Invented Universe: The Einstein- de Sitter Controversy (1916-17) and the Rise of Relativistic Cosmology*. Oxford: Clarendon Press.

Kidner, David W. 2014. The Conceptual Assassination of Wilderness. In *Keeping the Wild: Against the Domestication of Earth*, ed. George Wuerthner, Eileen Crist, and Tom Butler, 10–15. Washington: Island Press.

King, Amy M. 2010. Quietism and Narrative Stillness. *Common Knowledge* 16:532–551.

Kragh, Helge. 1990. *Dirac: A Scientific Biography*. Cambridge: Cambridge University Press.

Krauss, L. 2012. *A Universe from Nothing: Why There Is Something Rather Than Nothing. Afterword by Richard Dawkins*. New York: Free Press.

Krauss, L. 2017. *The Greatest Story Ever Told – So Far: Why Are We Here?* New York: Atria.

Kroupa, Pavel. 2012. The Dark Matter Crisis: Falsification of the Current Standard Model of Cosmology. *Publications of the Astronomical Society of Australia* 29:395–433.

Lai, Olivia. 2021. *The World's Mass Extinction Events, Explained*. Earth.Org. https://earth.org/what-and-when-were-the-mass-extinction-events/

Laino, Luigi. 2018. Is Knowledge of Physical Reality Still Kantian? Some Remarks About the Transcendental Character of Loop Quantum Gravity. *Found Physics* 48:783–802.

Lamm, Julia A. 2010. Romanticism and Pantheism. In *The Blackwell Companion to Nineteenth-Century Theology*, ed. D. Fergusson, 165–186. Chichester, UK: Wiley.

Lanza, Robert. 2007. A New Theory of the Universe. *The American Scholar [Essays]*:18–33.

Larmore, Charles. 2017. Hölderlin and Novalis. In *The Cambridge Companion to German Idealism*, ed. Karl Ameriks, 205–226. Cambridge: Cambridge University Press.

Law, John. 2004. And if the Global Were Small and Noncoherent? Method, Complexity, and the Baroque. *Environment and Planning D, Society and Space* 22:13–26.

Lazier, Benjamin. 2008. *God Interrupted: Heresy and the European Imagination Between the World Wars*. Princeton: Princeton University Press.

Lemaître, G. 1931. The Beginning of the World from the Point of View of Quantum Theory. *Nature* 127:706.

Levinson, Marjorie. 2007. A Motion and a Spirit: Romancing Spinoza. *Studies in Romanticism* 46:367–408, 559.

Liddle, Andrew, and Jon Loveday. 2009. *The Oxford Companion to Cosmology*. Oxford: Oxford University Press.

Livieri, P. 2019. *F.H. Jacobi's "On Divine Things and their Revelation": A Study and Translation*. PhD diss., McGill University. ProQuest.

Martin, Wayne M. 2007. Transcendental Philosophy and Atheism. *European Journal of Philosophy* 16:109–130.

Matthews, Bruce. 2014. The New Mythology: Romanticism Between Religion and Humanism. In *The Relevance of Romanticism: Essays on German Romantic Philosophy*, ed. Dalia Nassar, 203–217. Oxford: Oxford University Press.

Meillassoux, Quentin. 2011. *After Finitude: An Essay on the Necessity of Contingency*. Translated by Ray Brassier, Preface by Alain Badiou. London: Continuum.

Melamed, Yitzhak Y. 2000. On the Exact Science of Nonbeings: Spinoza's View of Mathematics. *Iyyun: The Jerusalem Philosophical Quarterly* 49:3–22.

Melia, F. 2013. The $R_h = ct$ Universe Without Inflation. *A & A* 553.

Mendelssohn, Moses. 2012. *Last Works*. Translated by Bruce Rosenstock. Champaign: University of Illinois Press.

Millán, Elizabeth. 2007. *Friedrich Schlegel and the Emergence of Romantic Philosophy*. Albany: State University of New York Press.

Millán-Zaibert, Elizabeth. 2005. The Revival of *Frühromantik* in the Anglophone World. *Philosophy Today* 49:88–108.

Moenkemeyer, Heinz. 1977. François Hemsterhuis. Admirers, Critics, Scholars. *Deutsche Vierteljahrsschrift Für Literaturwissenschaft Und Geistesgeschichte* 51:502–524.

Naess, Arne. 2016. *Ecology of Wisdom*, ed. A. Drengson and B. Devall. London: Penguin Random House UK.

Nassar, Dalia. 2011. 'Idealism Is Nothing But Genuine Empiricism': Novalis, Goethe, and the Ideal of Romantic Science. *Goethe Yearbook* XVIII:67–95.

Nassar, Dalia. 2012. Spinoza in Schelling's Early Conception of Intellectual Intuition. In *Spinoza and German Idealism*, ed. E. Förster and Y. Melamed, 136–155. Cambridge: Cambridge University Press.

Nassar, Dalia. 2014. *The Romantic Absolute: Being and Knowing in Early German Romantic Philosophy, 1795–1804.* Chicago: University of Chicago Press.

Nassar, Dalia. 2020. An 'Ethics for the Transition': Schelling's Critique of Negative Philosophy and Its Significance for Environmental Thought. In *Schelling's Philosophy: Freedom, Nature, and Systematicity*, ed. G. Anthony Bruno, 231–248. Oxford: Oxford University Press.

Neubauer, John. 1993. Science and Poetry. In *The New Princeton Encyclopedia of Poetry and Poetics*, ed. A. Preminger, T. V. F. Brogan, F. J. Warnke, O. B. Hardison Jr., and E. Miner, 1120–1127. Princeton: Princeton University Press.

Neurath, O. 1973. Wissenschaftliche Weltauffassung: Der Wiener Kreis. In Empiricism and Sociology, eds. M. Neurath and R. S. Cohen. *Vienna Circle Collection* Vol 1. Dordrecht: Springer. https://doi.org/10.1007/978-94-010-2525-6_9

Newton, Isaac. 1999. *The Principia: The Authoritative Translation: Mathematical Principles of Natural Philosophy.* Translated by Bernard Cohen, Anne Whitman, and Julia Budenz. Berkeley: University of California Press.

Newton, Isaac. 2014. *Newton: Philosophical Writings*, ed. A. Janiak. Cambridge: Cambridge University Press.

Novalis (Georg Friedrich Phillip von Hardenberg). 1891. *Novalis: His Life, Thoughts, and Works.* Ed. and Translated by M. J. Hope. Chicago: A. C. McClurg.

Novalis (Georg Friedrich Phillip von Hardenberg). 1997. *Philosophical Writings.* Ed. and Translated by M. M. Stoljar. Albany: State University of New York Press.

Novalis (Georg Friedrich Phillip von Hardenberg). 2007. *Notes for a Romantic Encyclopedia.* Ed. and Translated by David W. Wood. Albany: State University of New York Press.

Oerlemans, Onno. 2002. *Romanticism and the Materiality of Nature.* Toronto: University of Toronto Press.

Ogden, Mark R. 1989. Amor dei Intellectualis: Hölderlin, Spinoza and St. John. *Deutsche Vierteljahrsschrift für Literaturwissenschaft und Geistesgeschichte* 63:420–460.

Olbers, Heinrich W. M. 1826. On the Transparency of Space. *The Edinburgh New Philosophical Journal* 1:141–150. https://babel.hathitrust.org/cgi/pt?id=hvd.32044106376346;view=1up;seq=7

Ortega y Gasset, Jose. 1964. *What is Philosophy?* Translated by Mildred Adams. New York: W. W. Norton.

Otto, Rudolph. 1926. *The Idea of the Holy.* Translated by John W. Harvey. London: Humphrey Milford, Oxford University Press.

Owens, Thomas. 2015. *Nature*'s Motto: Wordsworth and the Macmillans. *Notes and Queries* 62:430–435.

Pagels, Heinz. 1982. *The Cosmic Code: Quantum Physics as the Language of Nature.* New York: Simon & Schuster.

Peacocke, Arthur. 2004. Articulating God's Presence in and to the World Unveiled by the Sciences. In *In Whom We Live and Move and Have Our Being: Panentheistic Reflections on God's Presence in a Scientific World*, ed. P. Clayton and A. Peacocke, 137–154. Grand Rapids: William B. Eerdmans Publishing Company.

Pethick, Stuart. 2015. *Affectivity and Philosophy after Nietzsche: Making Knowledge the Most Powerful Affect.* Basingstoke/New York: Palgrave Macmillan.

Pnevmonidou, Elena. 2011. Veiled Narratives: Novalis' *Heinrich von Ofterdingen* as a Staging of Orientalist Discourse. *German Quarterly* 84:21–40.

Poe, Edgar Allan. 1848. *Eureka: A Prose Poem.* New York: G.P. Putnam. https://onlinebooks.library.upenn.edu/webbin/gutbook/lookup?num=32037

Pojman, Paul. 2023. Ernst Mach. In *The Stanford Encyclopedia of Philosophy*, ed. Edward N. Zalta and Uri Nodelman. https://plato.stanford.edu/archives/win2023/entries/ernst-mach/

Raiger, Michael. 2010. Coleridge's Theory of Symbol and the Distinction Between Reason and Understanding: A Genealogical Recovery of the Baconian Method of Science. *History of European Ideas* 36:311–323.

Rees, Martin. 1999. *Just Six Numbers: The Deep Forces that Shape the Universe.* London: Phoenix.

Rees, Martin. 2018. *On the Future: Prospects for Humanity.* Princeton: Princeton University Press.

Rhook, Graeme, and Mark Zangari. 1994. Should We Believe in the Big Bang?: A Critique of the Integrity of Modern Cosmology. *PSA: Proceedings of the Biennial Meeting of the Philosophy of Science Association* 1:228–237.

Richards, Robert J. 2002. *The Romantic Conception of Life: Science and Philosophy in the Age of Goethe.* Chicago: University of Chicago Press.

Rigby, Kate. 2001. Recovering from the Fall: The Greening of Modernity. *Journal of the Australasian Universities Language and Literature Association* 96:35–48.

Rigby, Kate. 2004. *Topographies of the Sacred: The Poetics of Place in European Romanticism.* Charlottesville: University of Virginia Press.

Ripple, William J., Christopher Wolf, Thomas M. Newsome, Mauro Galetti, Mohammed Alamgir, Eileen Crist, Mahmoud I. Mahmoud, William F. Laurance, and 15,364 scientist signatories from 184 countries. 2017. World Scientists' Warning to Humanity: A Second Notice. *BioScience* 67:1026–1028.

Romizi, Donata. 2012. The Vienna Circle's 'Scientific World-Conception': Philosophy of Science in the Political Arena. *HOPOS: The Journal of the International Society for the History of Philosophy of Science* 2:205–242.

Rovelli, Carlo. 2016. *Reality Is Not what It Seems: The Journey to Quantum Gravity*. Translated by S. Carnell and E. Segre. New York: Riverhead Books.

Ryckman, Thomas. 2005. *The Reign of Relativity: Philosophy in physics 1915 – 1925*. Oxford: Oxford University Press.

Sachs, Aaron Jacob. 2004. *The Humboldt Current: Avant-garde Exploration and Environmental Thought in 19th-Century America*. ProQuest Dissertations Publishing.

Safranski, Rüdiger. 2014. *Romanticism: A German Affair*. Translated by Robert E. Goodwin. Evanston: Northwestern University Press.

Sandkaulen, Brigit. 2023. Jacobi and Spinoza. Translated by Joseph Carew. In *Friedrich Heinrich Jacobi and the Ends of the Enlightenment: Religion, Philosophy, and Reason at the Crux of Modernity*, ed. Alexander J. B. Hampton, 34–48. Cambridge: Cambridge University Press.

Schelling, Friedrich W. J. 1975. On the possibility of a form for all philosophy. Trans. Fritz Marti. *Metaphilosophy* 6:1–24.

Schelling, Friedrich W. J. 1989. *Philosophy of Art*. Ed. and Translated by Douglas W. Stott. Minneapolis: University of Minnesota Press.

Schelling, Friedrich W. J. 2001. *System of Transcendental Idealism*. Translated by Peter Heath. Introduction by Michael Vater. Charlottesville: University of Virginia Press.

Schelling, Friedrich W. J. 2004. *First Outline of a System of the Philosophy of Nature*. Translated by Keith Peterson. Albany: State University of New York Press.

Schelling, Friedrich, Friedrich Hölderlin, and Georg W. F. Hegel. 2003. Oldest System Programme of German Idealism. Translated by Andrew Bowie. In *Aesthetics and Subjectivity*, ed. Andrew Bowie, 334–335. Manchester: Manchester University Press.

Schimelpfenig, R. 2017. The Drama of the Anthropocene: Can Deep Ecology, Romanticism, and Renaissance Science Rebalance Nature and Culture? *Am J Econ Sociol* 76:821–1081.

Schlegel, Friedrich. 1982. *Excerpts from Dialogue on Poetry including the "Talk on Mythology"*. Translated by Ernst Behler and Roman Struc. In *German Romantic Criticism*, ed. A. Leslie Willson, 84–110. New York: Continuum.

Schlegel, Friedrich [Also contains some fragments by Schleiermacher, Novalis, and August Wilhelm Schlegel]. 1991. *Philosophical Fragments*. Translated by Peter Firchow. Foreword by Rodolph Gasché. Minneapolis: University of Minnesota Press.

Schleiermacher, F. 1996. *On Religion: Speeches to Its Cultured Despisers*. Translated by Richard Crouter. Cambridge: Cambridge University Press.

Scruton, Roger. 1995. *A Short History of Modern Philosophy: From Descartes to Wittgenstein*. London: Routledge.

Scruton, Roger. 2002. *Spinoza: A Very Short Introduction*. Oxford: Oxford University Press.

Short, T. L. 2011. Did Peirce Have a Cosmology? *The Transactions of the Charles S. Peirce Society* 46:521–543.

Silk, Joe. 2001. *The Big Bang*. New York: W. H. Freeman.

Smeenk, Christopher. 2014. Einstein's Role in the Creation of Relativistic Cosmology. In *The Cambridge Companion to Einstein*, ed. M. Janssen and C. Lehner, 228–269. Cambridge: Cambridge University Press.

Smith, John H. 2014. Friedrich Schlegel's romantic calculus. In *The Relevance of Romanticism: Essays on German Romantic philosophy*, ed. Dalia Nassar, 240–254. Oxford: Oxford University Press.

Smolin, Lee. 2015. Temporal Naturalism. *Studies in History and Philosophy of Modern Physics* 52:86–102.

Smolin, Lee, and Roberto Mangabeira Unger. 2015. *The Singular Universe and the Reality of Time: A Proposal in Natural Philosophy*. Cambridge: Cambridge University Press.

Smoot, George, and Keay Davidson. 1993. *Wrinkles in Time*. New York: Avon Books.

Snow, Katherine C. 2021. Transforming the finite: Toward a Jacobi-ian ecological critique of big bang cosmology's new German romantic philosophy of nature [Doctoral Thesis]. Edinburgh: The University of Edinburgh Research Archive (ERA).https://era.ed.ac.uk/items/73af798a-6724-4ae3-91d5-b1554b404753/full

Snow, Katherine, and Sjoerd Kluiving. 2022. Reading the Anthropocene Through Landscape Archeology: Historical and Contemporary Landscapes of Accumulation in the Netherlands and Lebanon. *The International Journal of Environmental Sustainability* 18:1–22.

Spinoza, Baruch. 2009 [1677]. *The Project Gutenberg E-Book of* The Ethics. https://www.gutenberg.org/files/3800/3800-h/3800-h.htm

Stahmer, Carl. 2005. *Romanticism, Hypertextuality, and Metavisual Information Theory*. Phd diss., University of California Santa Barbara. ProQuest.

Stannard, Russell. 2004. God in and Beyond Space and Time. In *In Whom We Live and Move and Have Our Being: Panentheistic Reflections on God's Presence in a Scientific World*, ed. P. Clayton and A. Peacocke, 109–120. Grand Rapids: William B. Eerdmans Publishing Company.

Stern, Robert, and Tony Cheng. 2023. Transcendental Arguments. In *The Stanford Encyclopedia of Philosophy*, ed. Edward N. Zalta and Uri Nodelman. https://plato.stanford.edu/archives/fall2023/entries/transcendental-arguments

Stevenson, Leslie. 2015. A Theory of Everything? Kant speaks to Stephen Hawking. In *Inspirations from Kant: Essays*, ed. L. Stevenson, 64–76. Oxford: Oxford University Press.

Stoeger, William R. 2010. The Big Bang, Quantum Cosmology and creatio ex nihilo. In *Creation and the God of Abraham*, ed. D. Burrell, C. Cogliati, J. Soskice, and W. Stoeger, 152–175. Cambridge: Cambridge University Press.

Stöltzner, Michael. 2015. The Logical Empiricists. In *The Oxford Handbook of Causation*, ed. Helen Beebee, 108–129. Oxford: Oxford University Press.

Stroe, Mihai A. 2014. Poe's Hieroglyphic Universe: The Master Keys. *Romanian Journal of Artistic Creativity* 2:19–81.

Tegmark, Max. 2007. The Multiverse Hierarchy. In *Universe or Multiverse?* ed. Bernard Carr, 99–125. Cambridge: Cambridge University Press.

Tegmark, Max. 2015. *Our Mathematical Universe*. London: Penguin Random House.

Theisen, Bianca. 1998. Romantic Myths of Myth: Myth as Autopoiesis. In *Myth and the Making of Modernity: The Problem of Grounding in Early Twentieth-Century Literature*, ed. Michael Bell and Peter Poellner, 9–23. Amsterdam and Atlanta: Rodopi.

Theisen, Bianca. 2003. χα Absolute Chaos: The early Romantic poetics of complex form. Trans. K. Barry. *SiR* 42:301–321.

Thielke, Peter Graham. 2010. Who's Who from Kant to Hegel II: Art and the Absolute. *Philosophy Compass* 5:398–411.

Thrower, James. 1980. *The Alternative Tradition: Religion and the rejection of religion in the ancient world*. The Hague: Mouton.

Thurneysen, Eduard. 1964. *Dostoevsky: A Theological Study*. Translated by Keith R. Crim. London: Epworth Press.

Tiercelin, Claudien. 1998. Peirce's Objective Idealism: A Defense. *Transactions of the Charles S. Peirce Society* 34:1–28.

Tryon, Edward P. 1973. Is the Universe a Vacuum Fluctuation? (Letter). *Nature* 246:396–397.

Voltaire. 1901. Candide; or, The Optimist. Translated by William F. Fleming. In *The Works of Voltaire: A Contemporary Version*, Vol. I, 61–281. Paris: E. R. DuMont.

Vrahimis, Andreas. 2015. Legacies of German Idealism: From the Great War to the Analytic/Continental Divide. *Parrhesia* 24:83–106.

Wanning, Berbeli. 2017. Poet and Philosopher: Novalis and Schelling on Nature and Matter. In *Ecological Thought in German Literature and Culture*, ed. G. Dürbeck, U. Stobbe, H. Zapf, and E. Zemanek, 43–62. Lanham: Lexington Books.

Weinberg, Steven. 1992. *Dreams of a Final Theory: The Search for the Fundamental Laws of Nature*. New York: Pantheon.

Weinberg, Steven. 1993. *The First Three Minutes: A Modern View of the Origin of the Universe*. New York: Basic Books.

Weller, Shane. 2011. *Modernism and Nihilism*. Basingstoke: Palgrave Macmillan.

Wheeler, John A. 1982. Bohr, Einstein, and the Strange Lesson of the Quantum. In *Mind in Nature: Nobel Conference XVII*, ed. Richard Q. Elvee, 1–30. San Francisco: Harper & Row.

Wheeler, John A. 1999. Information, Physics, Quantum: The Search for Links. In *Feynman and Computation: Exploring the Limits of Computers*, ed. A. J. G. Hey, 309–336. Reading, MA: Perseus.

Wilczek, Frank. 2015. *A Beautiful Question: Finding Nature's Deep Design*. New York: Penguin.

Wilson, Andrew D. 1993. Romantic Cosmology. In *Encyclopedia of Cosmology*, ed. N. S. Hetherington, 596–604. New York: Garland.

Wittgenstein, Ludwig. 1958. *Philosophical Investigations*. Translated by G. E. M. Anscombe. Englewood Cliffs: Prentice Hall.

Zakariya, Nasser. 2017. *A Final Story: Science, Myth & Beginnings*. Chicago: University of Chicago Press.

Ziche, Paul. 2012. Monist Philosophy of Science: Between Worldview and Scientific Meta-Reflection. In *Monism: Science, Philosophy, Religion, and the History of a Worldview*, ed. Todd Weir, 159–177. New York: Palgrave Macmillan.

Życiński, Józef M. (Zycinski, Joseph M.). 1987. The Anthropic Principle and Teleological Interpretations of Nature. *Review of Metaphysics* 41:317–333.

Życiński, Józef M. (Zycinski, Joseph M.). 2002. Metaphysical and Epistemological Presuppositions in Stephen Hawking's Interpretation of the Creation of the Universe. *Roczniki Filozoficzne (Annals of Philosophy)* 50:109–133.

Zylstra, Stephen. 2023. Spinoza, Emanation, and Formal Causation. *Journal of the History of Philosophy* 61:603–625.

Index[1]

[1] Note: Page numbers followed by 'n' refer to notes.

K. C. Snow, *Our Inevitable Third Spinoza Controversy*, https://doi.org/10.1007/978-3-032-03045-0

Kolibri 360 GmbH
Ferdinand-Jühlke-Straße 7
99095 Erfurt, Deutschland
produktsicherheit@kolibri360.de